AF609349

# from Coast & Cove

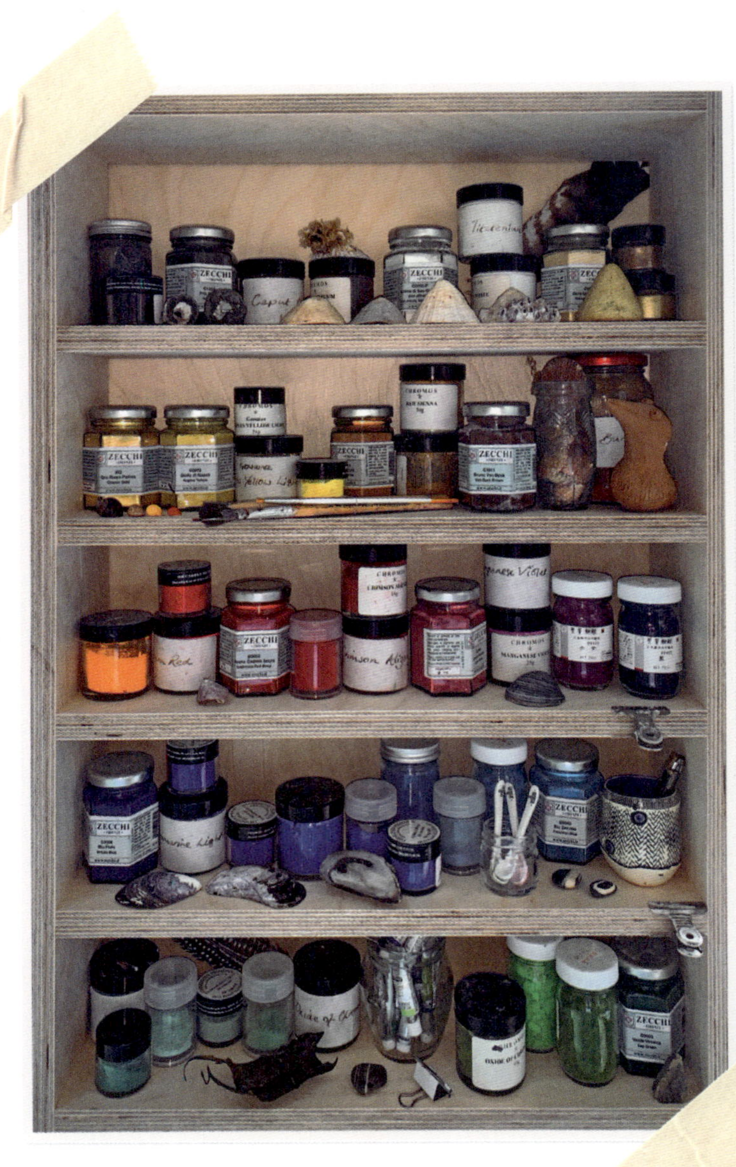
ZECCHI
CHROMOS
RAW SIENNA
Caput
Red
CHROMOS
MANGANESE VIOLET

# from Coast & Cove

## An artist's year in paint and pen

Anna Koska

PAVILION

*For Koska, Flora, Ellie and Jolly*

First published in the United Kingdom in 2022 by Pavilion
43 Great Ormond Street
London
WC1N 3HZ



ISBN 978-1-91168-212-7

A CIP catalogue record for this book is available from the British Library.

10 9 8 7 6 5 4 3 2 1

Reproduction by Rival Colour Ltd, UK
Printed and bound by Toppan Leefung Ltd, China

www.pavilionbooks.com

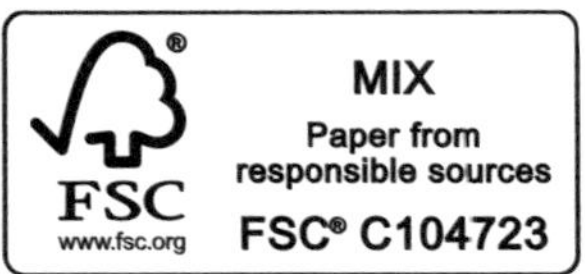

# Autumn

*21 September*
*Morning, rain-rinsed and lit.*

*Last boxes arriving today.*

*Been twenty years since we last moved, long enough to forget just how miserable it can be.*

*Nearly there though. A swim will go a long way to soothe both body and soul.*

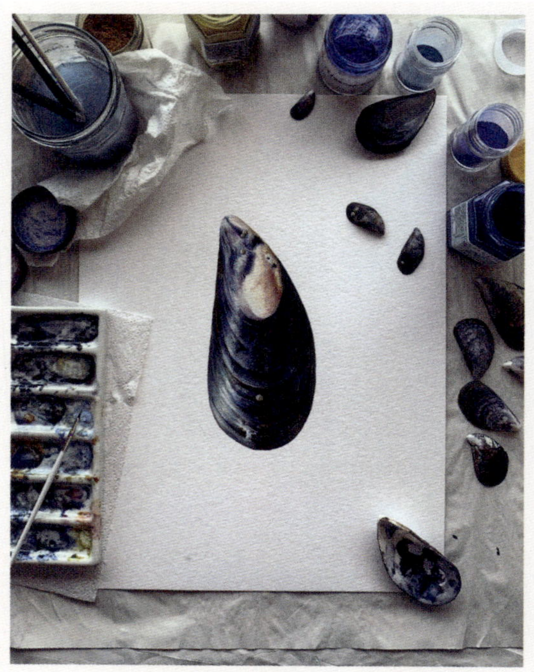

# The Mussel

Powdered pigments in their small, stout glass jars are laid out in front of me, their true colours muted, like parched puddles waiting for the rain to bring them to life...

cerulean blue
Vandyke brown
cobalt blue
oxide of chromium
perylene violet
ultramarine
sap green
Naples yellow

When I've stirred them through with a little raw egg yolk and water, they will glow once more with, hopefully, the same intensity as the mother rock from which they were mined. These colours reflect a change in both subject matter and mood.

I'm squaring up to paint a mussel shell.

Since I last wrote, we have upped sticks, and moved. My views are no longer held in check by a bank of old pine but now stretch out across folds of sheep-nuzzled fields, which lead one's gaze through their clefts and dips toward a new view: a jostling cluster of mostly whitewashed homes nestled around the edges of a tidal estuary. I can't see this stretch of water, but it's enough to know it's there. In my mind the thought of this move has been on the horizon for many years, though mostly as an indistinct yet hopeful dream. I never quite believed it would happen.

But these times are strange. The world is in the grip of a pandemic; just writing this sounds like the clichéd opening lines for a dystopian horror film set in the future. But the truth is, there is a virus trying very hard to survive and reproduce, and every country is endeavouring to counter and stem its flow.

And although there are times when I feel incredibly fearful for my family, for my friends, for every soul standing in its path, life goes on, whether we attend to it or not.

I'm dipping the tip of my brush into the small cup of egg yolk, now into the cobalt blue powder, now over to the white saucer where I gently mix this medium to create a smooth paste. A dab more water, now – breath held tight – I take this loaded brush over to the paper to slowly scribe the first gentle curving outline of the mussel.

I pocketed this shell while on a walk along the pebbled shoreline of a tiny little cove that seems only ever to be occupied by preening cormorants and the occasional seal.

It is worn, battle-scarred, but so well built by its creator that, judging by its size and thickness, it lived to quite an age. The life that attached itself to this mollusc is spattered and colourful, valiantly worn like some sort of Pollock fashion statement: the algae bloom of different hues and textures, the grid of teeny eggs laid by a limpet that believed this shell to be a safe ride. There are very neat holes bored into this shell... driven through by opportunistic whelks, hell-bent on a free snack.

And then there's the inside... the smooth, pearlescent cocoon, perfectly cupped to hold and protect the soft delicate body of the mussel inside. A little armoured haven.

It is quite beautiful.

To my eye, there's a dizzying array of 'treasure' to be found while beach roaming, but I always seem to gather at least one mussel shell, absent-mindedly slipping it into the

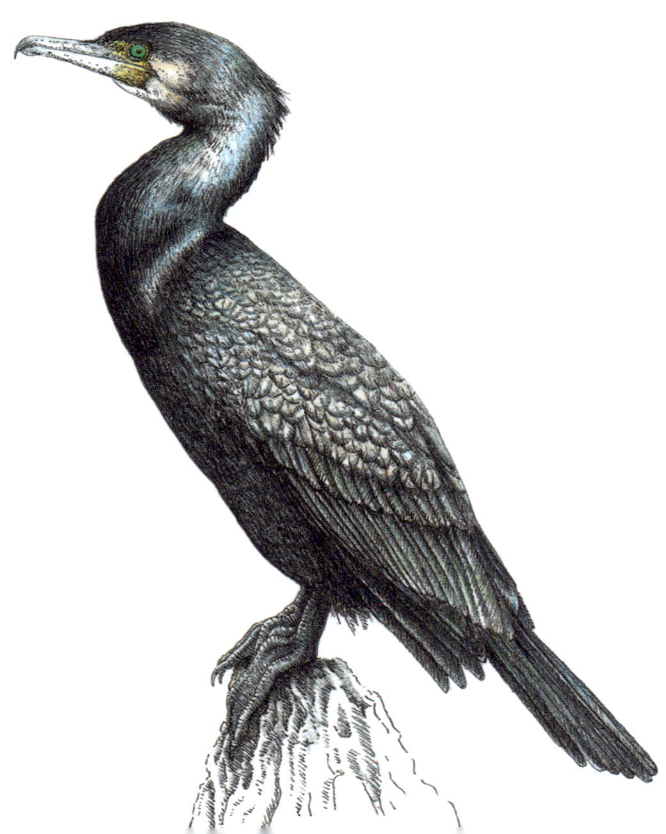

pocket of my coat, perfect for an idle hand to slowly trace the graceful butter-knife-sharp curve. The thumb, on its blind journey, will unfailingly catch on any tiny outcrop of micro barnacles and will find itself circling and returning to this little world. Then, without a thought, my hand will flip it prone side up, presenting a sensuous thumb slide, from the pallial line and up to the hook of the umbo. I think I could draw these shells with my eyes closed. At least in my head and my heart.

On reaching home, the contents of my pockets are usually washed and dried on a tray in the bottom of the oven. Then they're stored in Kilner jars to be admired and dipped into, when needed.

For the last thirty years I've been collecting such treasure, storing it in jars and popping the lids to smell the sea. And sometimes this sated my desire until my next visit. But mostly it made me ache with longing.

But now I'm here, and near enough that with an onshore breeze I can smell the ocean from our bedroom window. Near enough that I can cycle there to walk and rootle along the tideline, swim even.

Outline done, it's time to be brave and flood the paper with the myriad of blues that make up this tough yet elegant shell.

*27 September*

*Turn left at 'egg box' crossroads, then plummet past hedgerows wearing clusters of blackberries, a bristle of skeleton hogweed standing guard. Follow the narrow swoop of moss-carpeted lane to dip down to the field of harvest-ready maize. On the right the ground drops away, the road held in check by a wall of tightly stacked slate that snakes and curves with the steep field beyond. Sheep perch among its tussocks and terraces to pluck and tear at breakfast. Turn right, by the old stone roundel sporting a jaunty grass cap. Listen out for the raven couple that carve their sky above; you may catch a flash of cerulean blue as the jay hastily retreats for the cover (and larder) of his small oak tree.*

*I rarely remember directions, and sadly wasn't born with the inbuilt compass that many seem to have.*

*So, this is my way… new waypoints, new sights. The new way home.*

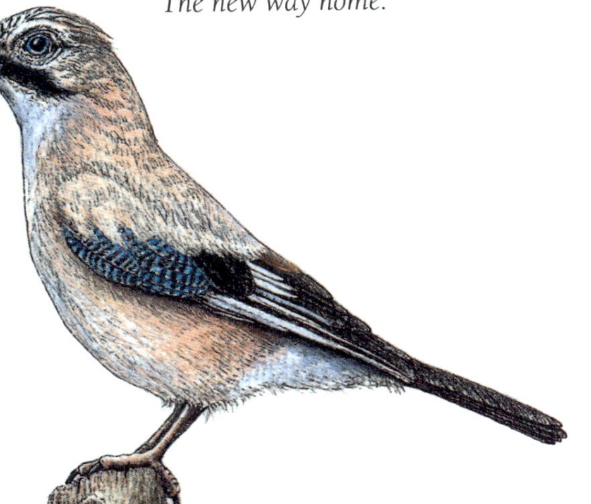

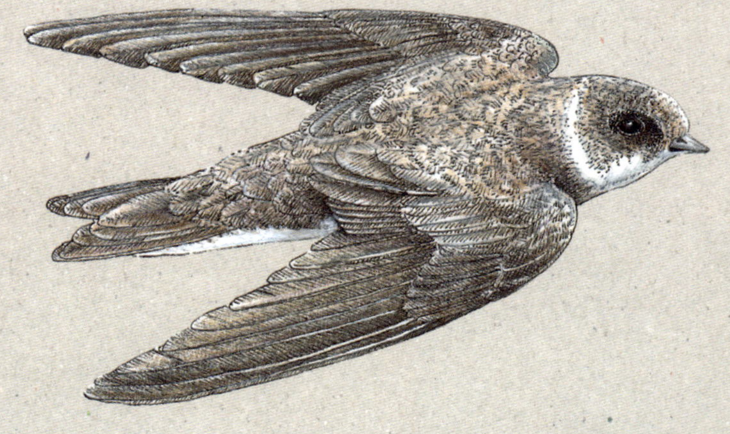

# Aviator

This morning we swam with sand martins.

It began with a bleary-eyed gathering of towel, costume and car keys as I ran for the back door. I'd got a message from our new neighbours that the tides might be 'right' and the weather kindly to introduce me to a new place to swim. I'm still working out the tides from the bristle of apps that I've enthusiastically downloaded to try to get my head around this part of the coast. But these two are like a walking, talking almanac, having lived here for decades. Nothing will come close to their almost visceral knowledge and feel for every inch of this part of Devon. If I'm lucky enough to reach old age, perhaps I too will have honed an eye and ear for such insights.

This cove sits snug and hidden, tucked into the base of a vertiginous drop of slate. It looks wholly inaccessible, but as you draw near it becomes apparent that it's an easy (if slightly precipitous) climb down, mostly made up of stone steps with an intermittent, looping, chained railing on the drop side, which lends one a sense of security, if only imagined.

Eighty-five steps and several kinks in the descent, and there we were, with next to no beach to stand on. The only sound to welcome us was the rattle and pull of the retreating sea as she combed her fingers through the multitude of stones that peppered the shore. No gulls, no cormorants. Not another soul. As we changed, a slow and steady mizzle began to fall, the kind that lands lightly on hair to make it at once appear cloaked in a diamantine veil. Last of the T-shirts and socks removed, and we were ready.

I looked up at last, steeling myself to start walking into the sea. It was only then that I took in the view. I realized that we were not alone. Against the smudges of sea and sky, a patrol of birds was busy carving up the air at such close quarters to the lilt of waves that I was certain they'd surely lose their aerial footing and hit the water. And yet, like circling acrobats, they fluttered and swooped with millimetres to spare. Fearful of disturbing their concentration, but keen for a better view, I decided to swim slowly out into the belly of the cove to see if I might be able to get closer. I hadn't considered that it might be them disturbing my progress, rather than the other way around; so confident were these pilots of their course that on several occasions I instinctively ducked to avoid what I thought to be a certain collision. Of course, it was incredibly arrogant to assume that a crash might have occurred, since it became patently clear who was the outsider here.

After a while, I stopped swimming to tread water and watch. Heart slowing, fingers and toes numbing. Their manoeuvres were at once loose then tight as they swung to and fro across the surface of the water. They were feeding. And I suspect the steady light drizzle made their quarry all the more easy to capture, given that every flying insect would be rendered a little heavier, and slower, by the moisture in the air.

Sand martins are the smallest of the hirundinids, with seemingly the weakest of flying styles, and yet they're the earliest to arrive in the UK, often touching down in March and settling into areas where the insects are most abundant. These would usually include lakes and reservoirs, river valleys and old water-filled gravel pits. But they're opportunistic feeders, willing to exploit whatever chances they can to find a good supply of food. This little cove sits just around the corner from the wide mouth of the estuary, with its silted banks of sand perfect for homemaking and rearing families. But we're now in early October, and these small birds had evidently hit upon a banquet and were feasting here to build up their strength and sustain them for their imminent departure for Africa. Usually a bird with a conversational, dry rasping stream of chatter, these sand martins were all but silent, intently focused on the job in hand.

As I continued to watch, freezing legs joining frozen arms in the race to chill my kidneys, they swooped nose-close to graze the rippled blue of the ebbing tide, and I forgot my discomfort entirely as one bird swept so near, I felt the very air it occupied move across my brow.

Breath taken.

Kidneys at last imprisoned in their very own ice hell, we swam for the shore and clambered out. As is often the case when in the midst of wild things going about their wild lives, I felt utterly ridiculous and incredibly clumsy, barely a cog in the grand scheme of things... and, like every human, probably more of a hindrance than a contributing part. But also awash with gratitude that I was able to be there and witness the ways of those wild things. We dressed. Somebody opened a flask and the smell of hot chocolate crept among us as half-filled, hot plastic beakers were passed around for each of us to hold and sip from. We spoke in hushed tones. To me the sound of a human voice might have felt irreverent, a spell breaker. I think the others felt the same.

Such moments as these. You could send me back to Sussex, or indeed ban me forever from returning to the sea. It would hurt. But I think I could just about cope, having stored this moment away among the files marked 'precious and unlikely ever to be repeated'. Though I dearly hope that one day we will be fortunate enough to witness these extraordinary aviators once more. Until then, I wish them fair weather and kind winds to carry them on their way back to Africa, where they'll overwinter and rest until next spring.

*6 October*

*Sitting in the car with Billie, waiting for a break in the weather so we can walk down to the sea.*

*Through the window is a flood of slate grey pouring into forest green; a shifting strip of pale teal uniting all.*

*Sky, tree, sea…*

*So much rain, still beautiful.*

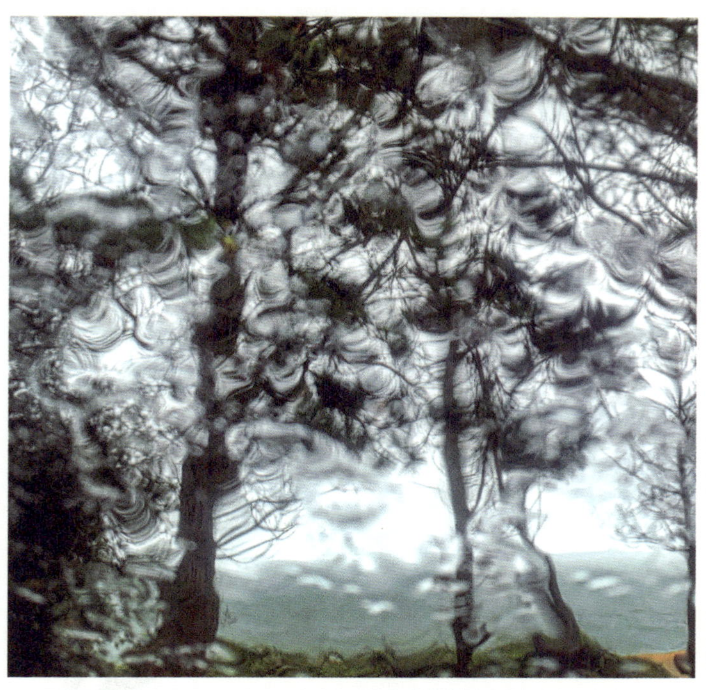

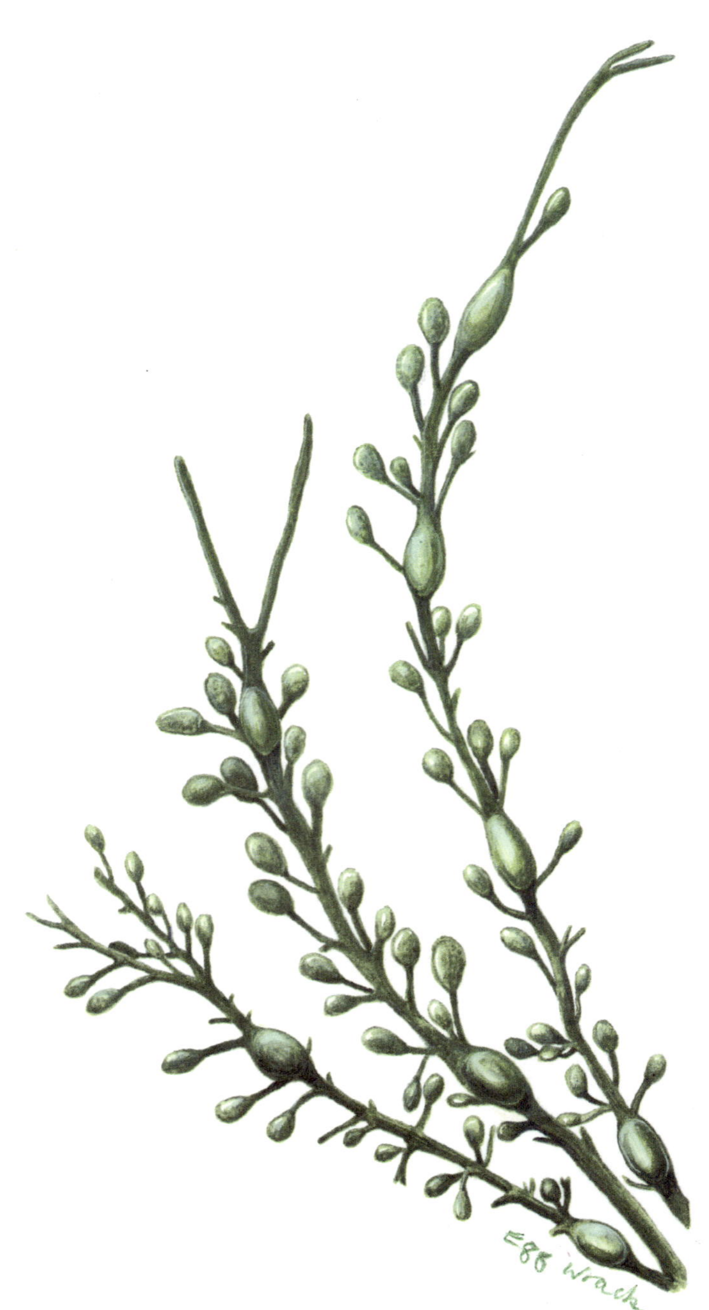

# Bones, Stones and Shells

A late afternoon walk with our lurcher, Billie, along a stretch of pebbled sand.

The sun is already ending her watch and heading west, skiving off in her eagerness to reach warmer climates. But in her wake, she's left a sky that is fast flooding to tangerine. And out in the bay, beyond the shadow cast by the land behind me, her likeness is mirrored in the ebbing tide.

I'm meant to be throwing the ball for Billie, but in truth both of us are distracted. She's new to sand, and seems surprised that a run for her ball takes a whole lot more effort, each paw-fall becoming lost among the shifting ground beneath her. She's stopped chasing and instead is working her nose through a heap of unanchored seaweed. It's a new perfume for her repertoire that used to be somewhat limited to fox, badger, deer and the occasional bird carcass. It's clear that this scent might become her current crush as the whites of her eyes are showing and her tail has begun its long-looping pendulum swing of excitement. Meanwhile, I've assumed my traditional ancient-heron stoop as I shuffle along the tideline in search of rare finds. It's one that my family is very familiar with, and it goes hand in hand with a tendency towards selective deafness because I'm incapable of stringing a thought – let alone a sentence – together while scanning for precious things.

I'd be lying if I said it was just mussels I pocketed. The truth is I'm one of those hoarders who gather up all kinds of detritus… bird beaks, feathers, crab carapaces and claws, sand-ravaged driftwood, dead beetles, shells, lesser-spotted dogfish egg sacs and finally stones. Stones of all kinds.

I think this fascination for beach rubbish began at about the age of eight when I went to stay with my grandparents in Mawnan, on the south coast of Cornwall. I would always be so excited for these weekend visits. But come the day, I'd be eaten up with a stomach churn of nerves. I always had huge misgivings of leaving the mother ship, and the idea of waking up in a strange bed in a very quiet household filled me with a sense of dread. But they were kind to me, and I suspect understood more than I realized; that I wasn't the most brave and confident child to have around. Having been bullied at school, and living quite remotely with very few children nearby, I'd become introverted and not particularly joyful about life. I think I became a bit of a personal mission for them and they mapped out my days with a fairly rigorous schedule of activities. On the first morning I awoke to a foghorn sounding. I climbed out of bed and walked over to my bedroom window, drawing back the curtains to find a gathering of cows emerging from a swirling sea fret, their brown eyes rolling as they strained over a stone wall to lick my window. The comedy of it all was not lost on me and I got the giggles. The bedroom door opened, and Grandpa crept in to see what was going on. He was dressed in pyjamas and a dressing gown. We chatted a bit while he gently helped me back into bed, explaining it was still a bit early to get up and if I could hold on just a little longer then we could all have breakfast together.

He left. I burst into floods of tears.

Within five minutes he was back again, still in his dressing gown, but this time holding a plate of something. I sat up in bed. And he sat down on the end. I realize now that Grandpa didn't know the first thing about cooking. But he knew how to make a sandwich, and most notably a brown-sugar-butter sandwich. I thought he was a genius. Unsurprisingly the day suddenly felt a whole lot brighter, and after breakfast we donned coats and boots, and headed out to do this thing called 'beach combing'.

Granny adored the sea, and one of her enduring pleasures, she said, was to walk the tideline and find the perfect white stone. It didn't have to be round, just one that 'fit' properly in her hand; one that could be reached for in a coat pocket, and held, rolled and smoothed by idle fingers. She called them worry stones and said that they were wonderful for easing troubled thoughts. So, lowering my head and hunching my shoulders to the weather I stalked along the beach beside her, in search of the 'right one'. I've still got it. It resides in a bowlful of cowries now, but for a long while it lived in my coat pocket.

Here, on this beach, entirely white stones are a rarity, but I've found another stone and it's green. Not just any green, but a mirage of many. I mentally flip to my pigments...

oxide of chromium, sap green and a dusting of *pria negra*, a deep graphite colour that I found in a tiny art shop in Florence. The stone is silken smooth, any edges and corners softened through many years of living life at the sharp end of geological history. I do the 'pocket test' to see if it fits well in my palm, if it's shaped to warrant rolling and if my thumb can find, and then rest comfortably on, one of its facets. It works.

I've been trying to learn a little more about where we now live. Bird books have new pages marked, simply because where we were once surrounded by fields and woodland, we now have a whole menagerie of coastal birds to explore. Plant books are tagged in anticipation of new delights to be found in hedgerows, edgelands of beaches and banks of rivers. And I've dug out a 1970s *Collins Guide to Animal Tracks* in the hope that I can work out with whom we're sharing our little corner of Devon. I've also found a very enthusiastic geology site and, having browsed it briefly, my eyes locked on to the details of a green rock called 'lower Devonian hornblende schist'. A bit of a mouthful to say, it's ancient, even by rock standards; weighing in at over 400 million years old, it is the oldest rock in Devon. Schists were formed and brought to their present position by mountain-building processes involving the collision of two continents. The collision of two continents! Doesn't this sound incredible? Sadly, this extraordinary fact was delivered during a geography lesson in a voice that held about as much excitement as the school

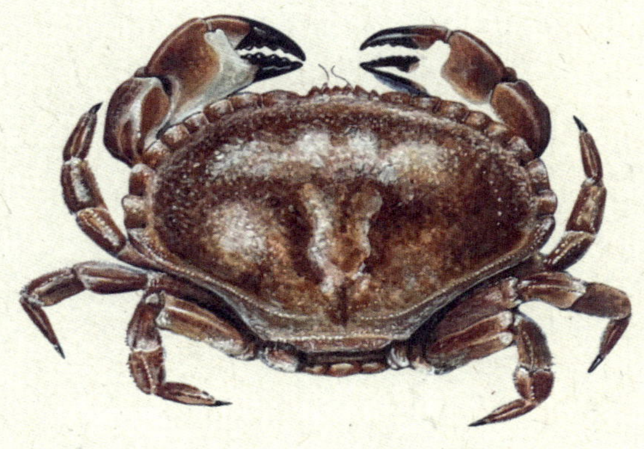

bus driver… 'Would you sit down? Take your feet off the seat in front of you.' I can remember drawing the tectonic plates and labelling the parts but feeling it to be about as believable as *Star Wars*. And a lot less exciting.

Smoothing this piece of stone in my chilled hands, watching Billie still working her nose along the seaweed tideline, it now feels relevant. I'm acutely aware that I've been fortunate enough to have spent the last fifty-four years happily anchored to this world. But I'm no more than a millionth of a blip in its history, and at some point, just like the crushed and morphed fossils contained within the minute silvered strati of this stone, I will become a mere speck among millions more that will make up the geology of this world in the future. A thought that would've completely panicked that shy, rather pensive eight-year-old, but one that gives me great comfort now.

Pocketing my new worry stone, I call Billie and we make our way back to the car.

*10 November*

*There are many quilted, iridescent balloons of beached Portuguese man-of-war to hopscotch around this morning. Over at the far end of the beach is a stranded young seal.*

*Kept Billie close by while the RSPCA gathered up the pup. Wouldn't have been a happy union!*

*Over the last few days storms have been raging, and yet the sea was a placid mirror for the fingers of sun that managed to slip past the barricade of clouds. Sea temperatures dropping fast… about 11 degrees today, according to my jam-making thermometer.*

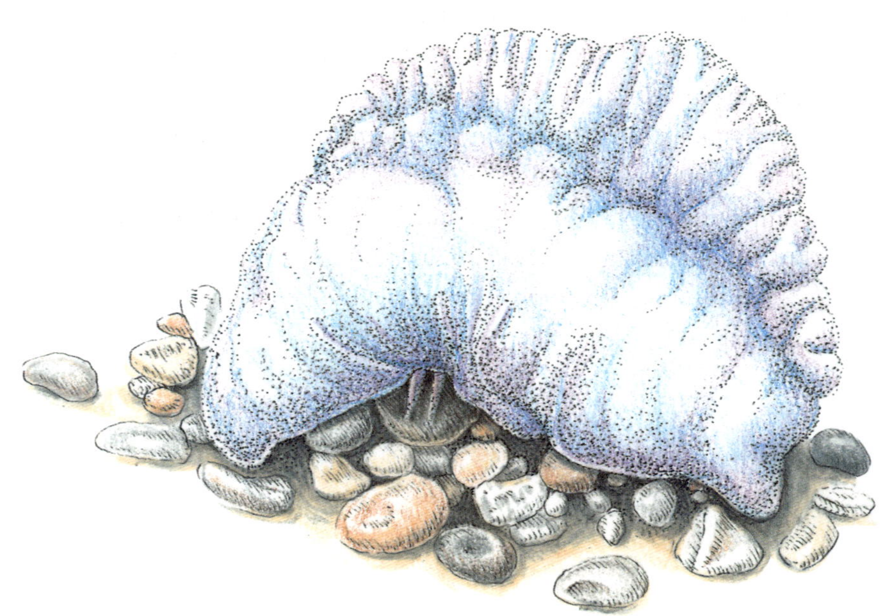

# The Texture of Water

I'm sitting in my studio, painting a hake while listening to *Clair de Lune* by Claude Debussy. It's a tender, sweeping piece that gently gathers momentum as it flows towards the main refrain, then slowly retreats to deposit the listener back at the beginning. With every swell of the piano, I can imagine the sway and shift of water.

And it takes me back to being perhaps thirteen and feeling more than a little awkward with the changes that were rewiring my body as I moved towards the springboard into adulthood. The ensuing few years were relatively mundane, punctuated by torturous bouts of self-loathing and a deep fear of being discovered to be a ghastly pubescent young woman.

We lived in Cornwall, near the sea, but with both of my parents busy working there was little time to visit the coast. However, every school holiday, our aunties and uncles would descend to stay nearby, and from then on, I could get lifts to the beach. And so many of my summers were spent armed with a towel and a swimming costume, hopping in and out of cars, and in and out of the sea. As an awkward teenager, brimming with insecurities, I didn't much like sunbathing. But being in the sea, in her cool embrace, I felt reborn as something far more agile and at ease. I felt cocooned and safe. I would stay in the water for hours at a time, just paddling about, diving under, propelling myself beneath the waves; bobbing up to wipe away the curtain of hair and, treading water, stare back at the beach.

The sea was, I think, my lifesaver. It brought me joy.

Until our move here, I'd almost forgotten the nigh-on magnetic pull that the sea had for me. Sure, I've swum in the sea since then but, with only the briefest window of opportunity that a holiday might offer, I didn't connect with it in the same way.

I've always been intrigued by the idea of swimming through the seasons, the bracingly brave and ever so slightly bonkers pursuit of wading in and voluntarily immersing one's body entirely, wearing little more than a few triangles of cloth. And having this stretch of water so tantalizingly close, it felt very much like a now-or-never moment in my life.

So, within days of arriving, and amid the mountains of boxes yet to be unpacked, I began to make regular trips to the beach to dip into the salty swell. Every time I swam the breath came easier, and the startled, reflexive lock of my diaphragm lessened.

It's November now, and the sea is colder than the body of water that swayed around these shores in September; any residual heat left over from the hot summer is long gone. But as well as being cooler, the very texture of the water has altered. It has become slower in its shift and roll as it sweeps in towards the shoreline of jostling pebbles, which brace themselves for the imminent grab and drag of each

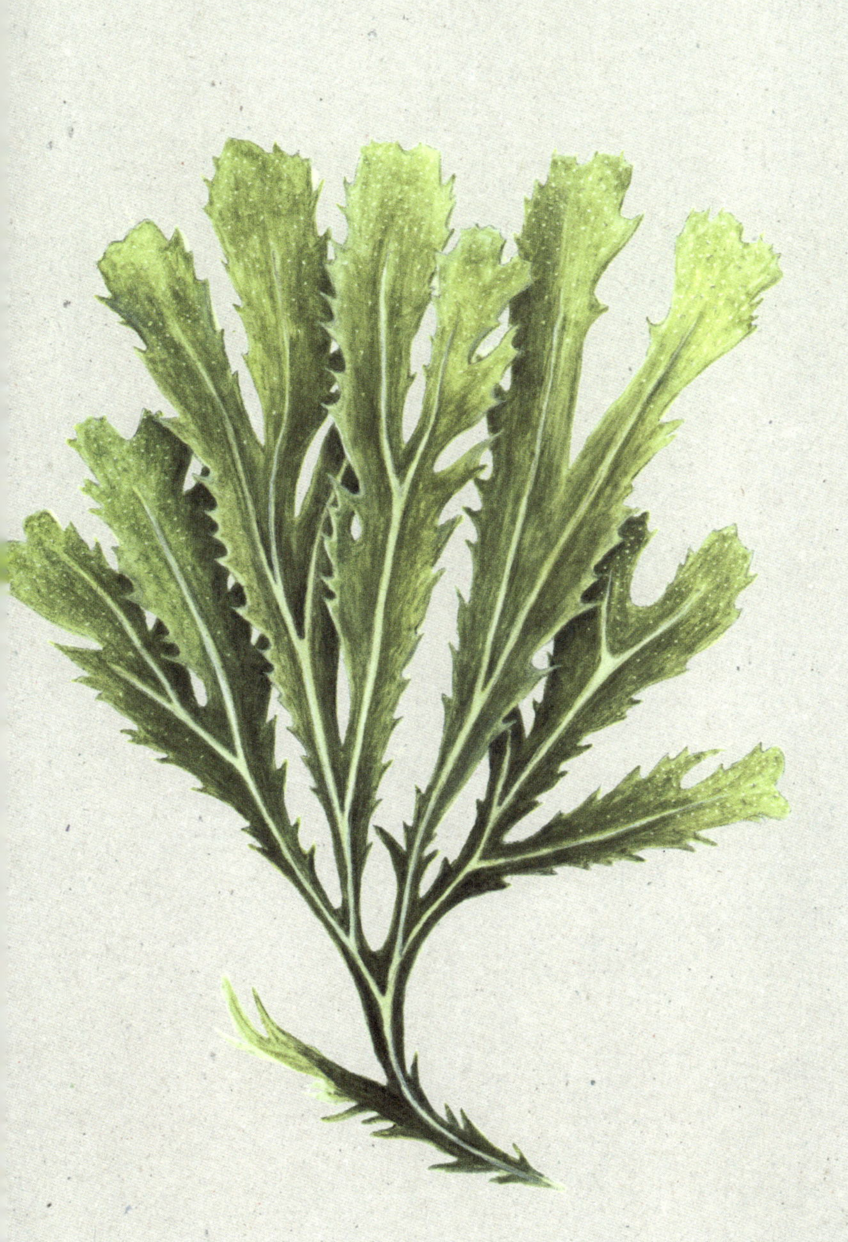

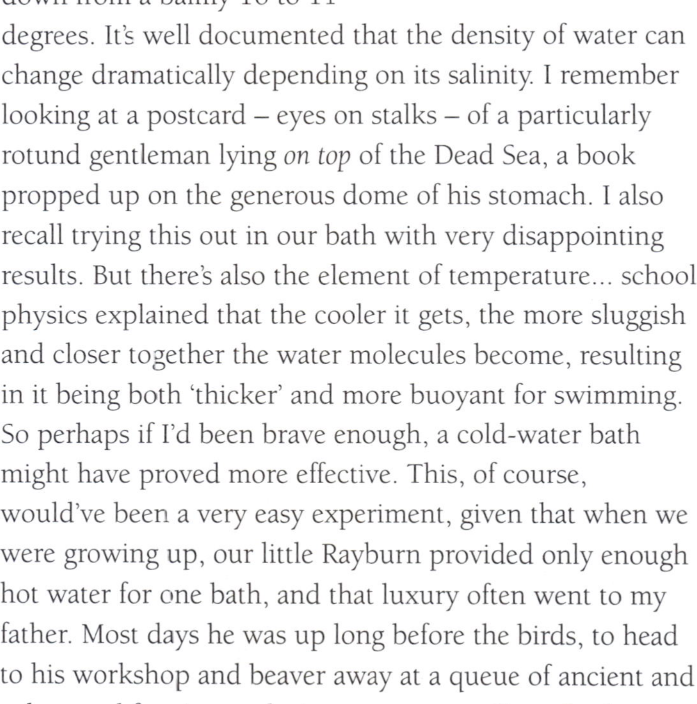

white frothing arm swinging in to pull at them.

In just two months we're down from a balmy 16 to 11 degrees. It's well documented that the density of water can change dramatically depending on its salinity. I remember looking at a postcard – eyes on stalks – of a particularly rotund gentleman lying *on top* of the Dead Sea, a book propped up on the generous dome of his stomach. I also recall trying this out in our bath with very disappointing results. But there's also the element of temperature... school physics explained that the cooler it gets, the more sluggish and closer together the water molecules become, resulting in it being both 'thicker' and more buoyant for swimming. So perhaps if I'd been brave enough, a cold-water bath might have proved more effective. This, of course, would've been a very easy experiment, given that when we were growing up, our little Rayburn provided only enough hot water for one bath, and that luxury often went to my father. Most days he was up long before the birds, to head to his workshop and beaver away at a queue of ancient and damaged furniture... laying veneer, mending a broken chair leg or reproducing a section of boxwood inlay for a sewing table. He didn't stop, and the queue of patients never seemed to shrink. My brother and I would be busy

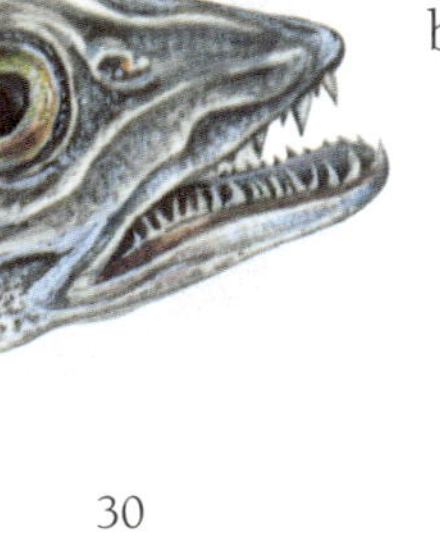

organizing our Weetabix into sections of Stonehenge when he'd appear at the kitchen door, ruddy faced, wearing a chest of wood shavings, his fingers coated in French polish. Brushing off the worst, he'd reach under the sink and pull out a tub of Nitromors (paint stripper). Opening this up he'd then proceed to pour it over his hands, and scrub. It would gnaw at his skin, and to manage the pain he'd mutter a stream of mostly incoherent swear words as he worked to remove the stains. The tap would be on full flow, pouring icy water on to his inflamed hands to soothe away the burn. Then all would become calm once more; hands dried, a mug of coffee in hand, face transformed to a wide grin, he'd head upstairs for his bath.

I swam this morning and the cold sea bit, and each thick furrow seemed to have more momentum and weight to it. Whereas the late summer sea seemed quicker to slice through, this water appeared keen to hinder my progress, with every stretch, cup and pull of an arm feeling more laboured and effortful. No longer silken, today it wore the quality of hessian, abrading my chilled limbs. But it felt real, cleansing, a relief. I suspect the grin I wore was not dissimilar to my father's.

And now, as I listen to this piece of music and plan tomorrow's escape to the sea, it's occurred to me that I've rediscovered a part of my childhood, one that gave me great comfort and left my mind rested. And in so doing, I've reclaimed a piece of myself that I'd somehow misplaced.

So here I am again, once more being held aloft by the sea.

# Winter

*12 November*

*The greylag goose* (Anser anser; *named by Swedish naturalist Carl Linnaeus and translating as 'think think').*

*A desire to find a higher point from which to view the curves and tributaries of the River Dart led us up and up the tiny unmetalled lanes that map out these parts. Then, on a whim, we took a sharp right through an open gate and into a grassy field. And, as we arrived, a dark cloud composed of more than three hundred greylag geese lifted as one and we stood there, jaws slack as these great birds, the largest in the goose family, flowed and rolled out across the land, calling out coordinates for the alternative feeding station, since we'd so rudely interrupted their breakfast.*

*I have never seen or heard anything like it.*

*As they peeled off and divided into smaller fleets to eventually disappear behind us, a new song filled the air, that of skylarks.*

*Skylarks! Here! Their bubbling song… it's the kind of music that has the power to lift the corners of a mouth and make the eyes shine.*

# A Thing about Fish

Yesterday's walk along the beach took my breath away for many reasons, not only because of the bitter easterlies that were tearing across the sands and whipping hair into our eyes but because Billie discovered a large dogfish. Also known as small-spotted catshark, or rockfish, these sleek and speckled members of the shark family live and feed along the sandy seabed of the shallower waters of the southwest. However, because of their willingness to scavenge for pretty much anything they can find, this often leads them into trouble with anglers, since they're much attracted to the offer of a free snack in the form of bait. There was a time when fish-and-chip shops used to serve them up, but they're now seen as having little commercial value, given that their rough ammonia-rich skin renders getting them to market a very labour-intensive process. So, sadly, they're mostly seen as a nuisance and often thrown overboard.

In the early 1500s, however, it was a different story, with the dogfish's abrasive skin being highly sought after. When dried and treated, the hard surface of dermal denticles (tiny, tooth-like structures) presented the perfect medium for furniture makers to smooth and polish wood. It's even been reported that microscopic particles found beneath the varnish on a 1711 Stradivari violin look remarkably like the remains of dogfish skin.

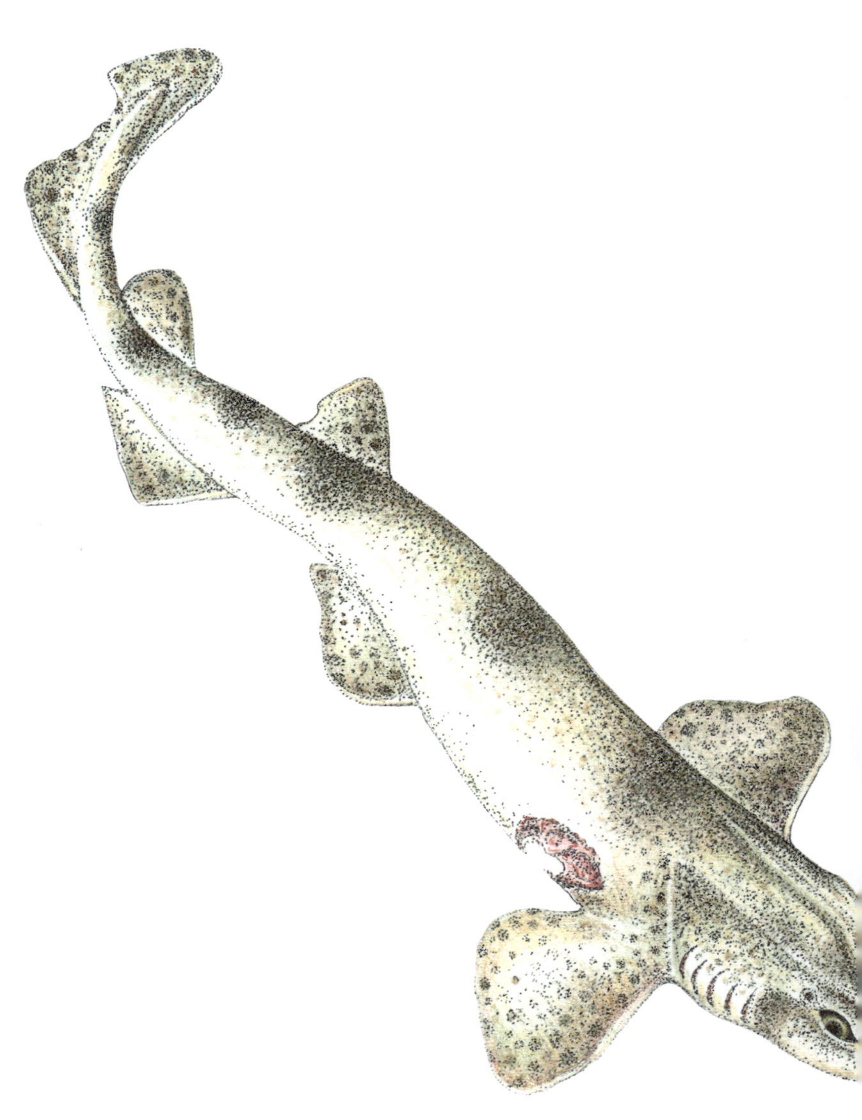

The poor creature lay belly up and prone on the pebbles, its milky white skin almost glowing in the slowly diminishing daylight. It was dead and, with a trauma puncture probably made by a hook, likely met its fate at the hands of a frustrated fisherman.

I was brought up on the north coast of Cornwall, and fish featured regularly on the menu at home. I'm not sure if we ever ate dogfish, but the first 'real' fish meal I can recall was conger-eel pie. By real, I mean it wasn't preformed to look inoffensive and un-fish-like.

Our little kitchen was the hub of familial life as we were growing up. Everything happened there. Various dogs carpeted the under-table area, waiting to hoover the inevitable fallout from above. There was usually a chair resting, legs up, on top of this table, enduring a 'French polish'. A sewing box took up a weekly slot, at the other end. The sink draining area doubled up as storage for clean pans as well as a resting place for a half-plucked goose or freshly gutted fish. The true heart of the kitchen, though, was the Rayburn, which sat quietly, always ready to cook or simply give comfort to a cold rear (unless the morning 'coal swing' had been forgotten).

I must've been about eight when, coming in from a prowl and play around the surrounding fields, I was welcomed by a distinctly different smell hitting my nostrils and throat. I walked over to the little stove and, on tipped toes, peered over the edge of Mum's vast and sturdy pot that always seemed to be bubbling on the single hob.

The bizarre sight of 10-centimetre-thick slabs of twitching muscle floating in a simmering sea of milk seemed entirely wrong to me.

'Mum! Stop cooking this thing! It's still alive!'

'Don't worry darling, it's really not.'

'What is "it"?' I asked, with not just a little misgiving.

'It's conger eel [bright, cheery voice], and [firm voice to discourage any possible revolt] it's going to be a wonderful fish pie.'

Not in the least reassured, I vowed that this was one dish that I wouldn't be eating, no matter how it was 'disguised'.

Of course, I did, and I absolutely loved it. Everything that was put in front of me I ate, not simply because I had to but because it usually tasted great. My dear mother was (and still is) is an extraordinarily inventive cook; one of those people who can always pull something out of the proverbial hat, even when the cupboard or fridge looks decidedly devoid of inspiration.

Living near the coast meant that our diet was rich in fish. It was cheap, too, really cheap. However, there weren't many English cookery books featuring fish recipes at the time, let alone a whole book dedicated to them. Consequently, Mum used to make it up as she went along. We ate a lot of sole, plaice and skate, and, of course, conger eel. But there was also mackerel.

I remember encountering my first mackerel at around the age of nine, when I pulled it out of the sea on a nylon line strung with hooks. It was all flex and flap, its tiger skin of azure and teal a shock as it danced in my sandcastle bucket. And I think it was at this point that I started to get a little obsessed with the sheer beauty of fish, any fish, even the so-called 'ugly' ones.

I still find the colours of a living mackerel to be quite the most vivid and hypnotic. But I've long since learnt that within hours of being caught, like all fish, its sheer brilliance becomes muted, a small act of retribution, perhaps, for hauling it out in the first place.

As an illustrator I work mostly from life and so it is when I draw or paint fish. But because of their tendency to fade, I've learnt to take many photos, and as soon as possible, to ensure my palette reflects the vibrancy of my model. I make a few sketches to work out proportions and composition. Then finally I bag it, label it and pop it in the freezer so I can refer to it when needed. So doing also means that I've a ready supply of fish to eat.

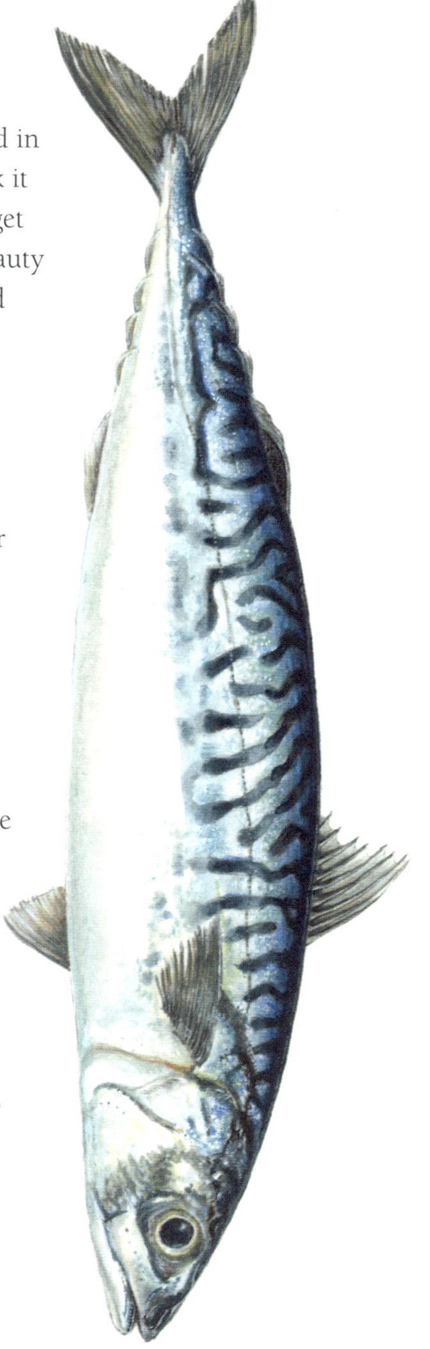

Some years ago, when I tagged myself predominantly as a food illustrator, I was asked if there were any guidelines I could share with anyone wishing to embark on a similar career. I thought on it and then penned this (perhaps slightly flippant) response; although it is essentially how I still approach most of my work.

GO AND FIND THE REAL THING – There are limits to this, of course. In the case of some obscure fish, I'm not often in a position to don a wetsuit and jump in the deep if it lives at the bottom of the inky blue, but there's always a fish market.

HANDLE THE THING – If it's got bones, discover where they lie. If it's got skin, get familiar with the feel and texture: hairy, bald, feathery, scaly, bit gooey, slippery, a bit want-to-put-it-down-and-run. Move it around. Check out how the shadows play on the dips and curves.

TAKE IT HOME TO MEET THE FAMILY – If I can buy it and fit it in the car then I will bring it back home. There I can do sketches and take stock photos of proportions, details and colours.

COLOURS CAN FADE – Photos (your own) are important for colour reference. Stock photos can help, but the liberal pressing of the 'enhance' button by some keen photo editor usually results in a distorted colour range! There are exceptions: Brussels sprouts really are fluorescent green.

YOU NEED A FREEZER – Chest ones are best. In among joints of unlabelled (they fell off) beasties, I have a few whole and named creatures, including the odd gurnard and mackerel. When I'm illustrating fish, or indeed anything that once walked, swam or pranced, I tend to pop them, bagged, in the freezer. I can then haul them out to check the odd detail if necessary.

PERFUME IS IMPORTANT – Choose a scent that you can stomach at close proximity. Anything zingy is good. Use this to spray the thing that once walked, swam or pranced; so doing stops the studio smelling like an A-level biology lab during dissection.

EXPERIENCE THE FOOD – You've got to have eaten the thing you're illustrating. Obviously. You can't expect to portray it as an edible, lust-worthy food if you've no sense of its taste, texture or soul.

KEEP YOUR DISTANCE – Keep 'helpful' people (not just children)/waggling paintbrushes/colour felt-tips/food-dipped appendages, at arm's length from your work.

Talking of hands, during my probationary period of illustrating I recall many traumatic moments of trying to remove handprints that had been concocted from a fabulous array of mixed media, including fish scales, lead pencil, watercolour, and the occasional smear of Malteser crumbs (added to create texture). From excessive hand washing I now enjoy the hands of a seventy-year-old (and paper margins that remain as white as bones), although they have guided me through my art and led me further afield than the dinner table to illustrate the wild and wider world beyond.

*10 December*

*Early morning dip, waiting for the sun to haul herself up. Just a thin, running stitch of silver for now, hemming the sky from the sea. Kept on grinning, still am. Have yet to get the hang of changing back into clothes more efficiently and with less 'beach' in my socks and knickers.*

*A friend reminded me of this repeating verse from A. A. Milne's poem, 'Sand-Between-The-Toes':*

*We had sand in the eyes and the ears and the nose,*
*And sand in the hair, and sand-between-the-toes.*
*Whenever a good nor'wester blows,*
*Christopher is certain of*
*Sand-between-the-toes.*

# The Clothes Press

Today has been wet, relentlessly so, and the rinsed air has carried more than a kidney-stab of cold. I've caught myself standing so stooped to keep warm that I swear I've just seen my dear (and long-departed) grandmother staring back at me from the mirror in our bedroom. Snapping upright I reach for the nearest jumper and hurry out.

Back downstairs, waiting for the kettle to boil. Beyond the kitchen window the saturnine sky is offering a stage to the trio of buzzards that have been calling to one another throughout this sodden day. They've been to-ing and fro-ing across the bottom of the garden between an old sycamore and a stand of stunted oak. Then, as if pulled by an invisible thread from above, their grey shadowy forms swoop out from the bare branches to climb the damp airways. But there are no thermals to play among today. Their ascension into the late afternoon is achieved through sheer muscle and determination. Having climbed up to what must be a better vantage point, they stay awhile, and then descend to respective perches. Within the hour they'll get itchy feet and do it all over again.

Without trying to anthropomorphize, it's hard to imagine that they enjoy a day of rain any more than we do. And they've still to find their next meal. But, judging by the fresh pile of wood-pigeon feathers I found this morning at the edge of a little copse, they're not doing too badly.

I say trio of buzzards, but in truth it's more likely to be a mated pair and an unhitched juvenile hanging around; the flying perhaps an attempt on the couple's part to defend their territory.

It's a thrill to find them here, given that we were lucky enough to enjoy their chatter almost daily where we lived before.

It's curious that when we upped sticks and moved here my biggest misgivings were bird-related. Not whether the wardrobe would fit in our new bedroom or if we'd find all our glasses reduced to a pile of shards at the bottom of the packing boxes. I panicked at the thought of no longer being able to see and hear our regular visitors. Over the years they'd not only become my signposts for the passage of each season but also my 'companions' who'd helped me take root and feel so wholly connected to the land there. Of course, birds don't do boundaries (at least not those made by humans) and there are plenty here. But I felt a tug of unease nonetheless… We left behind a large bubbling community of goldfinch, a song thrush with a repertoire that would make any seasoned chorister envious. There was a raven that would take his daily exercise among the clouds

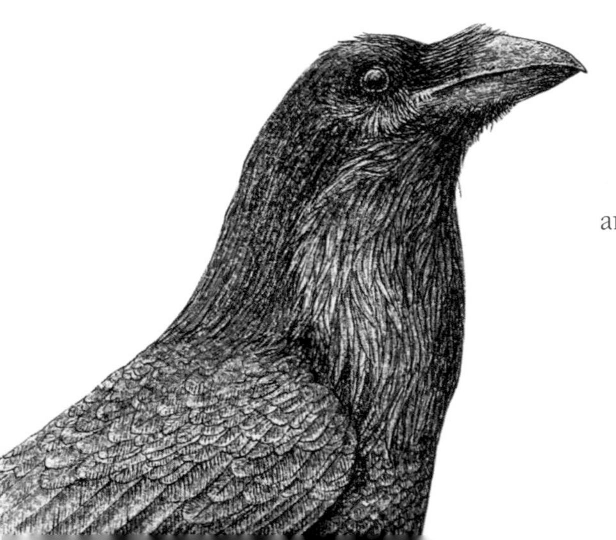

above our field every morning, his rattle-call announcing each lap. The Canada geese that would arrive with the first frosts of winter told me that it was time to fill the wheelbarrow with logs. And there was a wren that would thread my early summer mornings with a gilded stave of such enthusiasm that rising early became a mission rather than a moan.

I needn't have fretted, particularly where buzzards are concerned.

Since our arrival I've noticed that pretty much every tiny, windy lane I drive along has at least one buzzard playing at statues on a telegraph post, no doubt its chosen point from which to view its curtilage of feeding ground. They never move. They don't even bother to acknowledge my passing. The only living things that seem to disturb their concentration are members of the corvid family, usually a jackdaw, occasionally a raven. Then the mask of impassivity cracks and, with one deep wing push, they're off and scooping the air to move away from trouble.

Along with kestrels, crows, jackdaws, owls and sparrowhawks, buzzards were just one of the many patients that spent days and sometimes weeks at our house when I was a child. My father had a reputation as a fixer of birds (and other wildlife). Strangers would often turn up at our house holding a cardboard box, knock on the door, which was usually open, and out would come the familiar question:

'Are you Jeremy Bishop?'

To which my father would nod.

The contents of the box would be explained, and then finally revealed. Quite often there'd be a feeble explosion of feathers, at which point my father would quickly shut the flap and relieve the ambulance driver of their patient.

Dad explained to us very early on that the first thing that can kill a damaged bird is not the wound itself but the shock of the accident. So as a family we learnt just to carry on our lives around any cardboard box that appeared in the kitchen, usually next to the Rayburn, to keep the invalid warm. It was so tempting to open them, particularly if the patient became more perky and started to shift around inside its private room.

At some point Dad would quietly open the box and, gathering up the bird in his dry, callused hands, he'd have a closer look at the damage. I remember marvelling at talons, beaks, the steely, defiant gaze of the wounded, as if daring us to touch them, even if they were utterly incapable of self-defence.

Some were merely stunned from the bump of a car, others fearfully light from starving through an inability to feed themselves. These were often young fledglings found on the side of the road; perhaps their parent had been killed, or simply got spooked and left them there.

However, there were others, that were crumpled shadows. And these were the ones that would show on my father's face, with lines deeply etched.

Some never recovered.

But among the many that did I remember a rather skinny buzzard with a damaged wing. Every day he'd sit perched on the clothes press in the dining room, either preening or pretending to sleep. Very quickly he'd learnt that Dad was a source of food and so what began as disdainful distrust, rapidly morphed into actual excitement every time he entered the room. He gladly took pieces of raw meat, often roadkill pheasant. And whereas a previous kestrel patient preferred to have his food dragged tantalizingly across the floor so he could 'catch it', the buzzard was quite content to be almost spoon-fed. He started to gain some weight, and his wing began to mend. There were daily physio sessions, too. My father covered a length of wood in material, and gently nudged the buzzard at tummy height, so it would step on to the proffered perch. Then, by moving the stick up and down he would encourage the buzzard to use both of his wings to balance, so stretching and strengthening the damaged one in the process.

Then he'd be popped back on top of the clothes press from where he'd survey his domain. At that time, I was revising for my A levels, and the dining-room table was the only place I could spread out my books. My presence was not something that the buzzard was particularly bothered by and, except for the occasional rustle of feathers and shuffle of feet, I'd forget he was even there. However, the one time he did decide to remind me was when I had all my biology prep laid out, and he chose this moment to jettison a poo that flew two metres across the room and slapped on to my notes. Most who visited didn't believe he was real, though; my grandfather even enquired as to where my father had bought the stuffed bird from. 'So very lifelike.'

In time he healed well enough to try flight. After a few crashes and rather awkward retrievals, he made a full recovery and was released in a nearby field. He returned occasionally, as if to check in on my father, but eventually the visits became less frequent until he was entirely wild once more.

The day has now been swallowed up by the night. It's become cold, and I need to head out and fill a barrow with logs. 'Our' buzzards will have settled on to their preferred perches, and along the lanes I imagine that it's now the turn of the owls, who'll take on the evening shift.

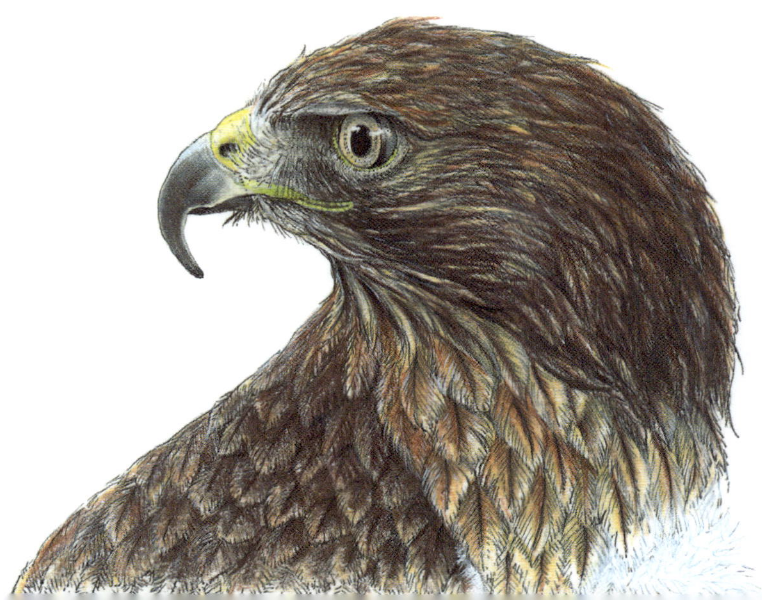

*25 December*

*Christmas Day swim with my family...*

*Shouting warm wishes to other families gathered.*

*Hands clasping hot chocolate, munching on mince pies and ginger biscuits.*

*Everyone distanced, but all of us in good spirits, perhaps united by the relief of feeling the sun on our faces; more than likely though the shared adrenaline kick of doing something ridiculous.*

*One of the strangest and potentially toughest Christmases that many have experienced. Saying a prayer that we all manage to find a little peace and joy during these few days. Here's to a kinder, calmer year ahead.*

# Curlew

Sunday, 3 January and the sun crested the hill behind our house, just after 8am. These winter mornings are dawning incrementally earlier each day, and today it was as if the clouds stood back as she entered stage left. She glowed. After a hurried breakfast, we packed up the car and, after tying the kayaks on the roof, my husband ferried me and our eldest, Flo, to Totnes. The plan was to paddle the 14 kilometres up to Dartmouth, bathed all the while in glorious sunshine. But by mid-morning the clouds had become less enchanted with the sun and, as we left the slipway in Totnes, the last wink of winter blue was snuffed out.

The clouds followed us all the way, a miserable troop of grey as we paddled along the Dart. But rather than tainting our journey with their reproachful gaze, they painted the most dramatic of backgrounds against which everything shone as if burnished. We passed by sleeping boats with their mooring lines swung and made taut by the ebbing tide; cormorants in full funeral garb sat poised and preening on their bobbing anchor buoys. They barely registered our passing; we were no more than a minor irritant to their otherwise peaceful day. Over to our right, oystercatchers pottered about on red, spindled legs, working their way along the slowly expanding muddy banks, intermittently stabbing for cockles and lugworms.

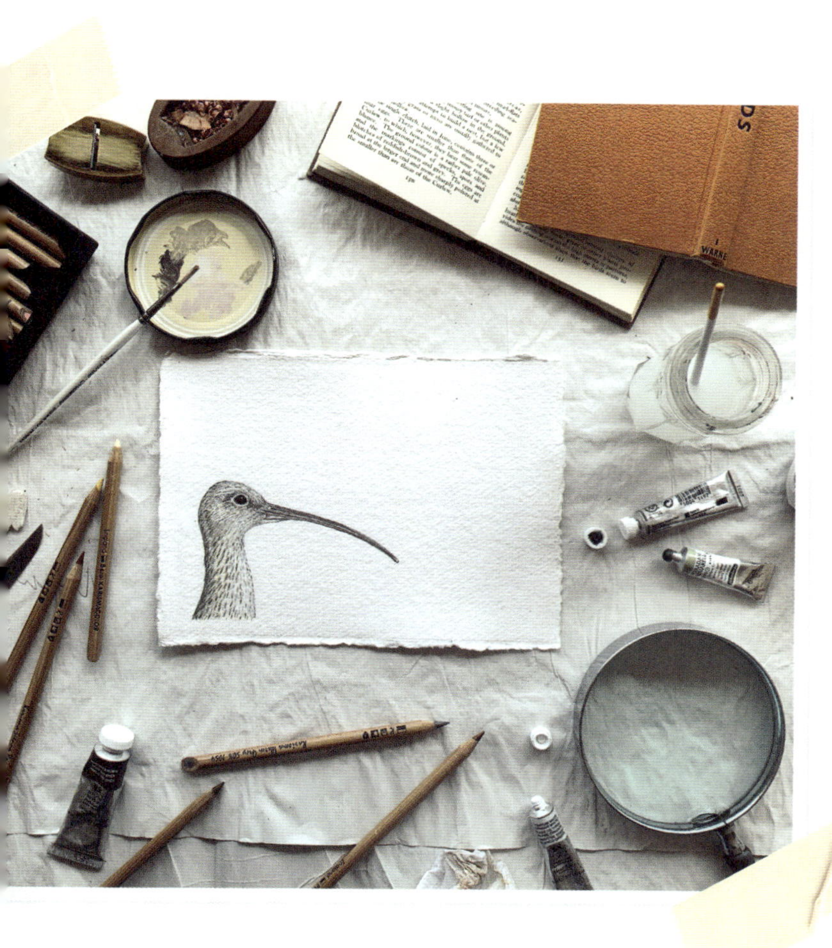

A little further on, past tightly packed banks of bulrushes, we forked left into a broader section of the slow-moving river and caught sight of the brightest streak of azure as a kingfisher jetted across the wind-ruffled waters to the safety of the tree-lined margins. I've never seen one before but had been tipped off to look out for them here, and recognize them by their direct style of flight. This one sliced the air at such speed I half expected a sonic boom!

The retreating tide was beginning to slow, and our paddling became a little more effortful. We managed to maintain a steady pace though and, as the river widened once more, we spotted a family of seals spread out and snoring across a pontoon, which was listing heavily, presumably under their combined weight. Perhaps disturbed by our presence, one began to inch itself along and flop noisily into the river. It was the kind of casual manoeuvre one might make into a cool pool on a hot day: deliberate, and deliciously lazy. It surfaced briefly, its big, dog-like eyes blinking at us, then slipped beneath the breeze-scuffed ripples. The other seals remained supine and sprawled. It was Sunday after all.

Everything about this paddle felt charmed. And to be enjoying it with my daughter felt precious. No matter the frown of sky above, we felt buoyant and deeply content as we made our way along the slow, curving swell of water.

I paused a while, closed my eyes too, to hear the moment better.

It was then that my ears became snagged on a new sound, one that ranged high above the constant descant of gulls and crows.

A sweet, lilting cry slipped across the water to pluck at the hairs on the nape of my neck. It lingered as if suspended on the air, then faded. Three more calls.Then silence.

I've only ever heard recordings before.

Goosebumps now.

I looked around, eyes straining to search among the meandering stretch of gulls and oystercatchers, cursing that I'd forgotten to bring my binoculars with me. And then there it was, along with perhaps a dozen more, striding across the flats… curlews! Funny how once you've seen one, you can see more. They'd been there all the time.

We stopped paddling and let the tide carry us as we watched them rootle and feed. We had the best seats in the house. All that was missing was the comforting, authoritative voice of David Attenborough expounding on the characterful lives of these elusive birds. However, the moment was somewhat interrupted when we became aware that our kayaks had gently bottomed out on the muddy bank. With rigorous, whole-body wiggling we eventually loosened up and began to drift into deeper waters.

Conscious of the need to keep going to avoid paddling against the tide, we determined to step up our pace when one of the group took flight, right over me. I doubt I barely made a blip on its radar, other than perhaps as a flightless, flailing obstruction to clear, but it was all I could do to contain the bubbling shout of surprise and joy that threatened to escape. Everything seemed to slow as I looked up...

Things that registered:

The long, gracefully decurved beak, perfectly adapted for pulling out worms from the mudflats;

His eyes focused forward, not below at us gaping in wonder;

A neck and upper breast marked like a gathering flock of arrows, streaming then parting over his ivory belly to splay out in lined formation across the creamy underside of his wings;

Feet elegantly pointed as a principal dancer might, mid sauté;

The air that contained him, now empty.

I rested my paddle on the kayak, took a moment, eventually letting fly a few whoops into that empty space. I then began babbling excitedly to Flo, who smiled and nodded, taking it all in her stride... apparently, it's not unusual to see me like this.

I've friends who live near to curlew sites, either those frequented for overwintering or those nesting places that are faithfully returned to each spring for breeding. And I've envied them for this. For me this bird has held an almost mythical status, not so much for its looks but for the pitch and dip of its unforgettable call. I've marvelled at their song when watching nature programmes. I have relied on and rejoiced in the words of Heaney, Hughes and Yeats, all of them conjuring forth the bird's keening, watery echo and sculpting it to suit their point and purpose. But these have now become somewhat dull and vague impressions of something so vital and unique in its ability to pierce air, ear and heart so completely.

It's been a week since our trip. Work has taken over. Other than a couple of walks by the sea, I've barely left the spare room that I'm using as my studio.

But I can't get the call out of my head.

I need to go back to hear and see them again.

Perhaps I can find someone who can take me by land to a hideout to watch them, learn about them and sketch them.

*23 January*

*Water 6.6 degrees … giving, buoyant, exhilarating.*

*If you'd told me that the future me would*
*enjoy, even look forward to, swimming*
*in such temperatures, I'd have*
*laughed and said only the very*
*hearty and slightly unhinged*
*would opt for so doing over*
*a deep sofa, the weekend*
*papers and a mug of tea.*
*Clearly, I am ever so*
*slightly unhinged;*
*but happily so.*

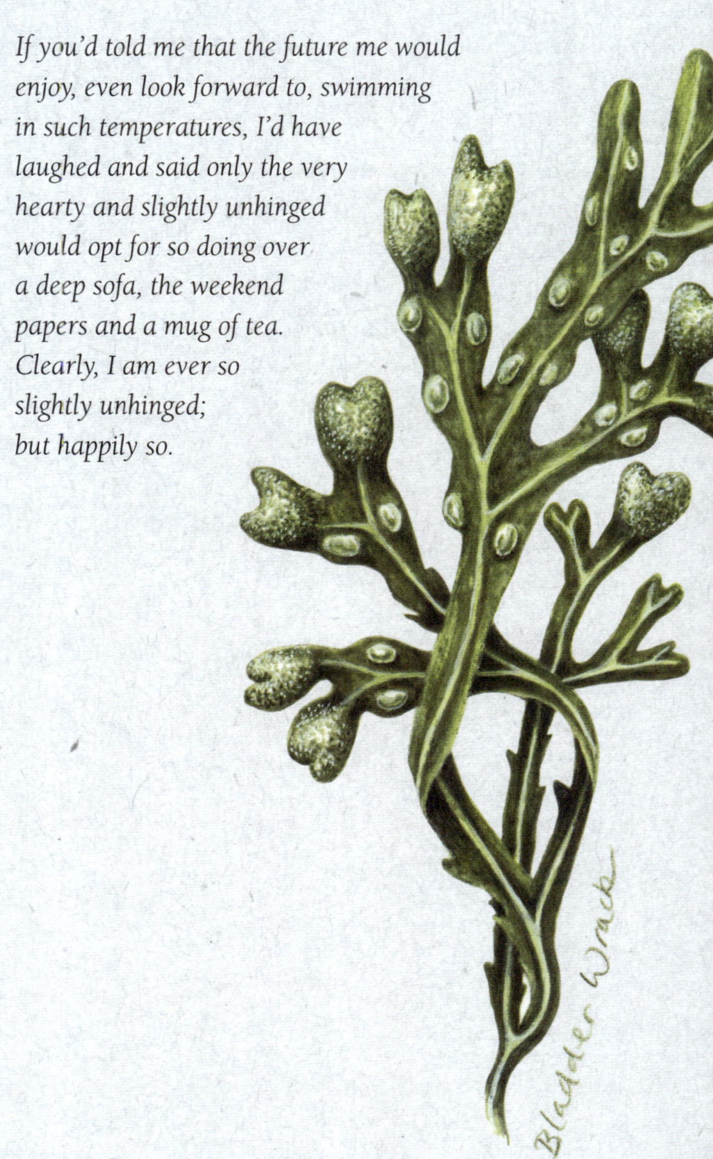

# God Bless Mrs Griffiths

Swimming in the sea has been a challenge at times.

Yes, the mercury is inching its way down my old thermometer, but also the stretch of beach from where I usually swim has a sudden drop away after the first few steps. When the sea becomes a little livelier this drop can play into the hand of an incoming tide. Waves gather momentum and can dump an unsuspecting swimmer back on the pebble beach, now sporting a new and ruddy rash from exiting the impromptu 'washing machine'. And if that hasn't got them, there's the inherent drag on their legs as they stagger out thinking they've cleared the water, only to be unfooted and face-planted by a sly retreating wave. Of course, I speak with authority on the subject, having enjoyed my fair share of giggles at the mercy of this sea.

For the last couple of weeks, the ocean has excelled herself, forming fleeting and intense alliances with passing storms. Together they've been sculpting titanic watery bodies that climb the air to dizzying heights before collapsing on to the battered shore with such pounding force that the beach reverberates to their drum.

On days like these, along with bringing a ball for Billie to catch, I've taken to stuffing my pockets with bags, because with every storm, the tideline becomes swollen with the untethered and unwary: treasure. While the winter-weary land provides little in the way of colour, the sea continues to give of herself, offering up the most delectable palette of hues in the form of seaweed. These colours range from the subtlest of blushes to hot-flush pink, from bottle green to parched sepia.

I've been searching for books about seaweed, and there's a chance I may need a new shelf for the slow but steady collection I'm gathering. Having lived inland for many years, my knowledge of this kind of flora is lacking, but I'm keen to learn. And from what I've unearthed so far, there is so much more to seaweed than meets the eye.

My earliest recollection of seaweed is of a dampish sprawl that appeared one morning, hanging from a butcher's hook outside our back door. My mother had been given this long and knotted specimen of bladderwrack by my uncle, who'd assured her that it was a better forecaster of weather than any weatherman on the television. At the time, I never understood just how this might work. To me, it looked like a sad and stranded sea creature. But it did seem a good predictor of rain. And perhaps there is a grain of working science to add weight to the story. When the air becomes damp before the onset of rain, the hygroscopic nature of the salt held within the structure of the seaweed naturally absorbs the moisture and renders the plant wet and flexible once more. This certainly seemed to be the case where ours was concerned. It would droop just a little lower, its dimpled air sacs miraculously reinflated. The seaweed lore then goes on to say that if the specimen becomes crisp and lifeless then dry and sunny weather is on the way. I suspect

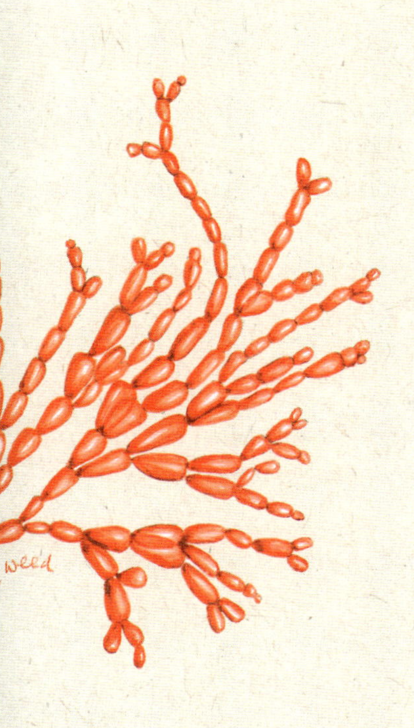
weed

that for many, predicting a rain-free day by studying a piece of dead seaweed might be pushing it. In fairness though, the meteorologists are often no more reliable when it comes to forecasting sunshine. But beyond the rather dubious weather divining, I found it fun to burst the little pillows of air.

From then onward, any family trip to the beach couldn't be considered a success unless I'd found some bladderwrack to squeeze and pop. I learnt that the best seaweeds were to be found casually draped over intimate gatherings of large rocks, which together formed perfect valleys for the intertidal pools. A variety of seaweeds would surround these pools, with their dipped tips swaying gently. A curious child reaching out an arm and sweeping aside the tresses might catch a glimpse of strawberry anemones, with tentacles in full bloom like the open face of a chrysanthemum. Transparent and sketch-like, tiny brown shrimp would pedal furiously for the safety of a shaded nook, and if the treasure hunter was lucky, a swarm of silvered, lesser sand eels might swirl and dart across the small, salty oasis.

I realize now that for me, those rock pools were not only charmed portals into another world but equally an escape from an otherwise rather dull one offered by grown-ups, who were more content to flop down on to the beach, chat, read and sleep. Sleep! Anathema to me, given that I'd already slept, all night! As an adult and parent, I can, of course, now appreciate why they might have needed to top up their reserves.

Today, as the winds are high once more, I'm picking over the bulky, lazy zigzag of debris to find specimens to bring home, label, press and paint.

The identification of seaweed doesn't come easily to me, particularly when the specimens are no longer suspended in the sea. So, I've learnt to fill up an old enamelled roasting tin with water and submerge the seaweed in it. Each strand can then become wet, buoyant and spread its various appendages to approximate its original glorious self once more.

The common names of seaweed vary from the rather unimaginative yet visually correct – fan weed, clawed fork weed, winged weed – to the completely fabulous: bunny ears, furbelows, creeping tongue weed, beautiful eyelash weed. And then there's a group that have given cause for great merriment, all prefixed with 'Mrs Griffiths's': Mrs Griffiths's fan weed, Mrs Griffiths's little flower, Mrs Griffiths's hairy basket weed. They always reduce me to childish giggles.

Mrs Amelia Griffiths was quite an extraordinary lady based here in Devon who, in the early 1800s, dedicated much of her life to collecting and identifying seaweed, and became responsible for raising awareness of the diversity of marine plant life in the area. Her reputation as a seaweed sage was even deemed sufficiently notable for the eminent Swedish botanist Carl Adolph Agardh to name a whole genus of red seaweeds '*Griffithsia*' in her honour. The honour was a rarity, since women were seldom, if

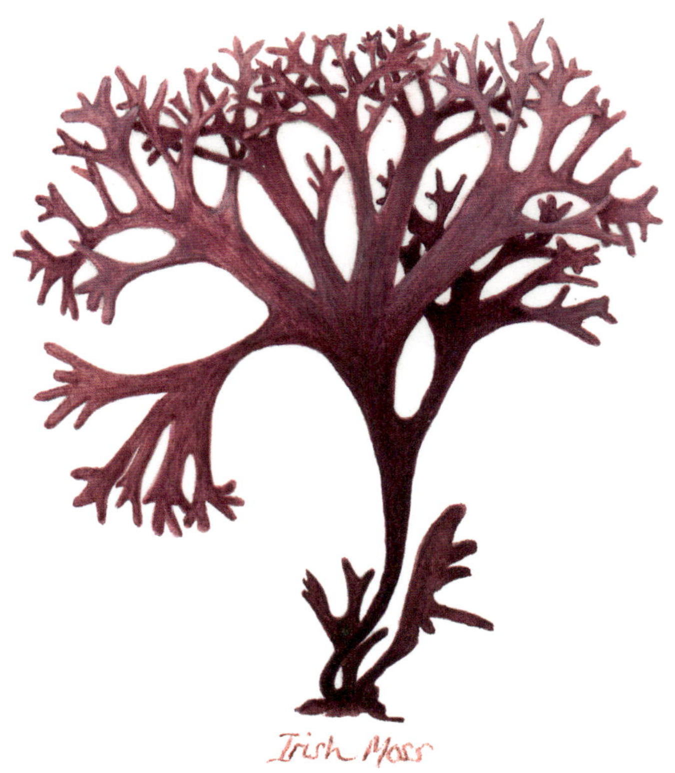

Irish Moss

ever, recognized for their research and contributions towards anything that might encroach on the largely male-dominated territory of 'science'. Wholly appropriate then that he felt it important to Latinize her name!

Home now, and I'm sipping a welcome mug of tea while the soaking seaweeds – Irish moss weed and an exquisite bloom of Mrs Griffiths's coral weed – quietly unfurl and inflate before my eyes. I will only ever be a very amateur botanist, but I am an avid enthusiast of this new branch of flora. And I'm itching to get my paints out and portray them.

Back to the bladderwrack… and of course we have some now, nailed up outside our kitchen window, and it's remained determinedly soggy.

Remarkably accurate, given that it has rained almost relentlessly for the last month.

23 February

*Tree on the edge of the garden has exploded… buds and blooms so thick, a freak drift of snow from nowhere.*

*Unsure whether it's blackthorn* (Prunus spinosa) *or its close relative, wild cherry* (Prunus avium).

*Ah, apparently there's one key difference. When the flowers are open, the small green sepals beneath lie flat against the petal if it's blackthorn. If, however, they're reflexed (bent backwards away from the petal) then it's wild cherry.*

*It seems this beautiful broad tree is a wild cherry. Can't wait for the bumblebees to find out.*

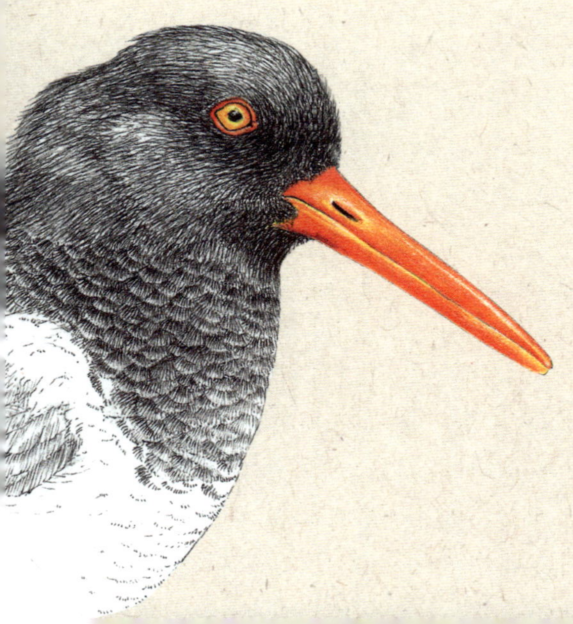

# Skin

It's been a long day, mostly spent hunched over an illustration of an oystercatcher. Shoulders aching and eyes craving a blast of fresh air, I grab for my coat and, thrusting my feet into the nearest pair of boots, I stomp out of the back door and into the last throes of a southwesterly. My neck tortoises back into its seasonal cradle of raised shoulders, and my eyes begin to tear up from the blessed cool.

It's late February now and the nights seem to have lost just a little of their raw edge, for the moment at least. Heading up the old farm track that wraps around the back of our house, I can just make out the clusters of snowdrops that have begun to appear along the steep bank of the field above. They've a soft glow about them like small, earthbound clouds caught in the fleeting gaze of the moon. I fumble with my torch for a better look. Each island is tightly packed with blooms, wimpled heads bent in prayer, or sleep. These early flowers feel like something of a miracle, given that only a week ago there was nothing here but winter.

The gentle breeze has snagged the tip of a sweet chestnut leaf. It is no longer an elegant sweep of acid green edged with teeth but a rusted, half-decayed shadow of last summer. This leaf worked hard to support its tree. I look up, seeking out the owner, but the dark mops up the last particles of light before they reach her.

In recent years it's become increasingly popular to believe that trees might possess a level of intelligence, a sentient quality, emotions. It wasn't so long ago, however, that being fond of trees, or even endeavouring to understand them or protect them was seen as a decidedly eccentric pursuit. Those who did choose to form close alliances – friendships even – with trees were seen as soft in the head, and not worth taking seriously. I have my father to thank for my slightly hippieish leanings. He works with wood (still does, at the age of eighty-six) to create beautiful furniture. He coppices woods to encourage growth. He cares for damaged giants, and respectfully fells the ones beyond mending, always with a mind to not waste any part of it; to make its passing matter. I just took it a stage further, always feeling compelled to get as near as possible to trees, to touch them, to marvel in their quiet magnificence.

A couple of days ago, the rain decided to take a break from its relentless assault, and the sun worked her small fingers into hairline cracks to rent great gapes in the clouds and flood the sky with light. So, we grabbed our coats, and Billie, and headed out.

There was one tree that I'd spotted on my way to and from the shops and had become a little obsessed with. It was a tall Scots pine that stood alone on the edge of a steep and deeply cleaved valley. Even though we knew that the path would be precipitous and muddy at best, none of us chose the right footwear, and so much of the descent was spent with our knees bent, zigzagging, slipping and sliding our way down to the valley floor. A busy, chattery brook divided the east- and west-facing slopes, with the tree rooted on the east. Jumping across I made the slow, steep ascent to meet the pine.

She stood perhaps fifteen metres tall and, with all but her very top branches snapped or torn off, she looked forlorn and quite vulnerable. Yet standing beneath this giant I couldn't for the life of me imagine her being any less strong, or beautiful, for the loss of these limbs. I leaned in to touch her. Set together like badly grouted crazy paving, a mass of thick, mouse-brown slabs formed her bark. Her dry, weather-beaten skin felt tired and ancient, but gave the impression of hiding something very much alive and poised for movement, as if she was simply waiting for spring so that she might flex, twist and shrug off this gnarled mantle to reveal her true and vibrant self.

With her roots dug so deep, it seemed that she surely had no feet at all. Yet they were most definitely there, spread wide beneath the valley; holding on tight since the day she'd decided to make this grassy gorge her home. No other tree had endeavoured to join her over the passage of time. She owned this valley outright, having endured perhaps 120 years of seasons and seas.

As a reward for her perseverance, she had uninterrupted views right down to a tiny bay, where the sea played hide and seek with the smallest crescent of beach.

Having at last met her, and feeling beyond lucky to have stood so close, I stepped back to admire her once more, and then turned to make my way down to that stretch of sand. I stripped off and walked into the sea and, once I'd cleared the first break, I turned to look back at her from this new vantage point. She seemed noble, a living beacon for all those who over the years might have sailed past this frayed stretch of coast.

I'm now looking down at the sweet chestnut leaf in my hand. I can see that the gradual deterioration of its once-glossy exterior has stripped it back, laying bare the remains of its venous system. It looks rather like the half-plastered, half-exposed workings of a wattle-and-daub wall.

The leaf confirms for me that in this world of trees and people we are more often alike than not. We're all endeavouring to grow and thrive, trusting that our chosen form, whether bark and leaf or bone and skin, will support us.

I'm looking at the skin of a leaf and I could just as well be looking at the skin of my hand.

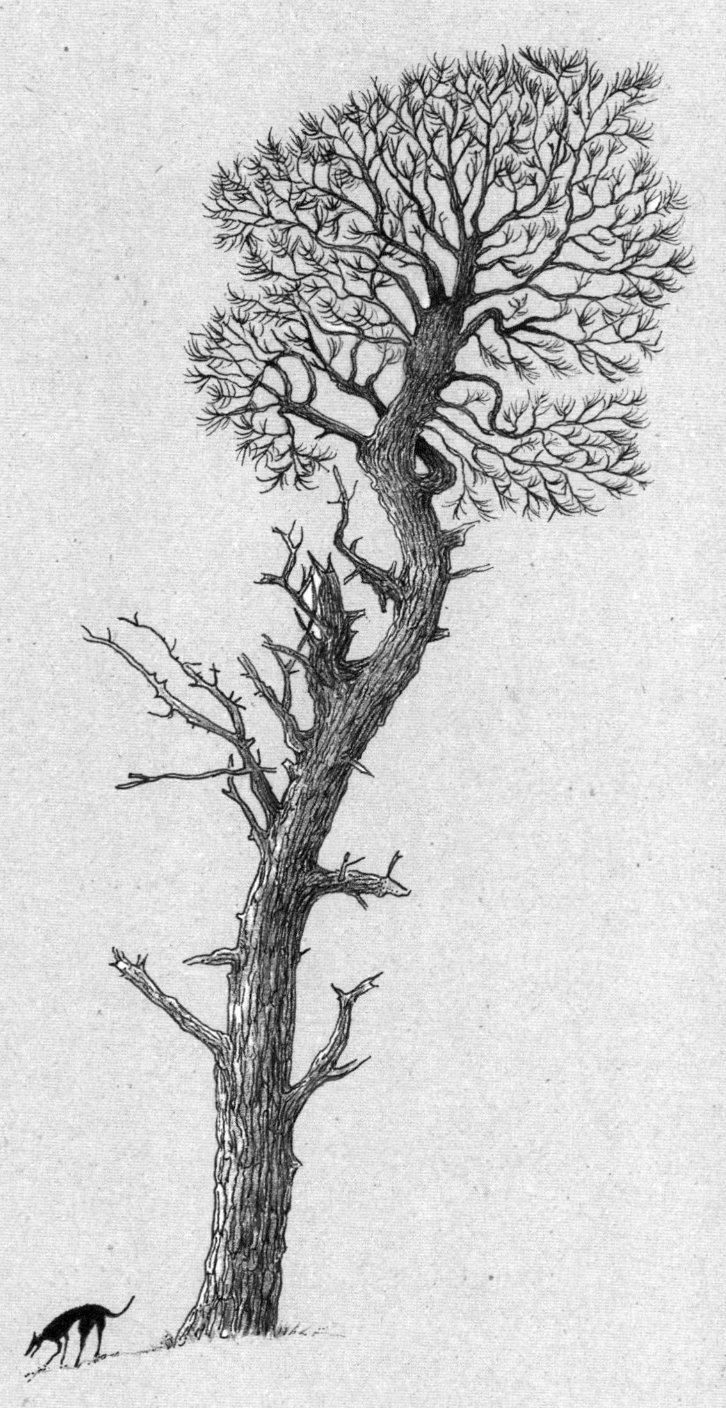

# Spring

*7 March*

*Dressing gown and wellies on, I'm picking early spears of wild garlic. Nearby, a buzzard has just swooped down as if to nail his breakfast to the grass. Moving on now to perch in an old ash, presumably to pick over his catch. As the temperatures begin to climb, the field mice have become more visible, scuttling among the flower beds and racing for safety from the ever-inquisitive Billie and her long, probing snout! It's little wonder that we regularly have three buzzards here. They have such convenient fast food.*

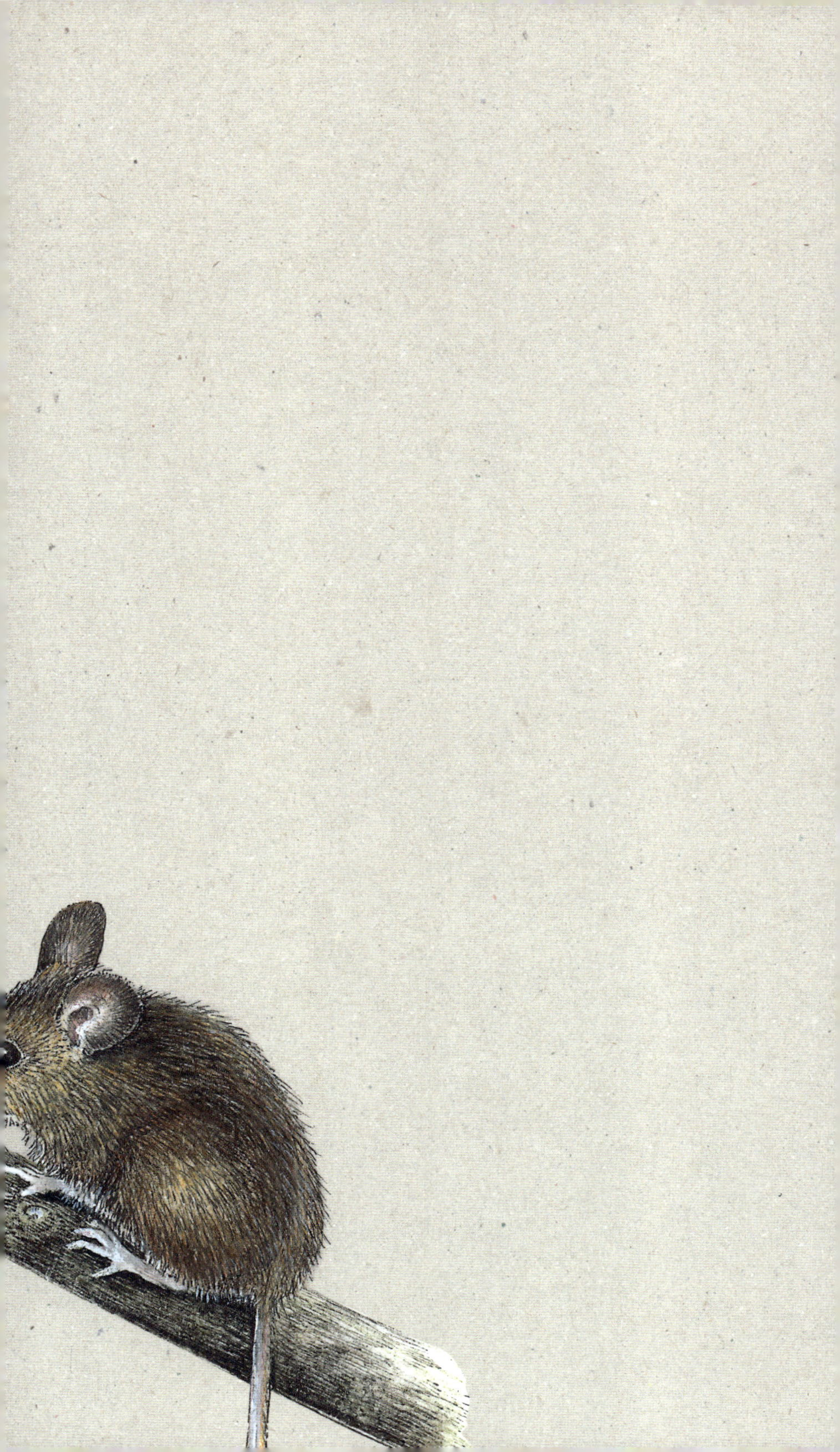

# Train Tracks and Clouds

There's the smallest sifting of light, just enough to see my feet. However, the squeak and crunch of frost beneath my boots is enough to guide me along the now-familiar path from our back door and out on to the grassy slope beyond. Every soft spear has become rigid, held in cold confinement. It'll remain thus until the sun breaches the high hill behind our house and slowly draws her warm hand across our land, releasing it from the tight grip of late winter.

I'm making my way down to the pond.

A few nights ago I took a stroll along the lane behind our house. I'd brought a torch with me and floated the

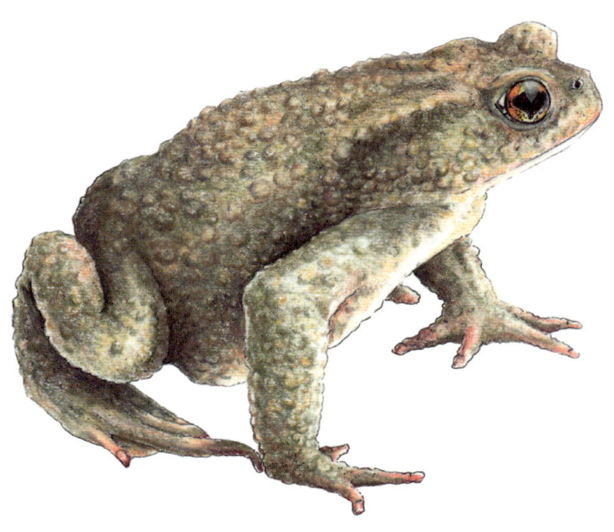

beam across the ruts and dips to avoid catching my toe and landing face down in the rain-filled puddles. Breath slowing, shoulders easing from a day at my desk. Everything was going well until my light found a male toad sitting bolt upright and entirely motionless, as if in a trance. I slowly crept up to him and squatted down to have a closer look, feeling sure that he'd move. But he didn't. Instead, he continued to stare into the middle dark, completely ignoring the intrusive torchlight and the invasion of his personal space. It was only when I moved the light around that I saw another toad, by my left foot. I'd so nearly stepped on him! Hastily rising, I decided to sweep the torch around just to make sure I hadn't squashed any on my way. I was stunned to make out a whole army of them, hundreds of eyes, lit and winking back at me like a line of low-hung fairy lights. Quite miraculously, it seemed I'd missed every single one. So many toads! Enough to reconsider the veracity of all those raining animal stories!

Taking in this astonishing scene, I could see that while some were also playing at statues, others were ambling very slowly but with purpose. And I realized it was in the vague direction of our pond.

I walked further along the track, hopscotching around little clusters here and there. One of the females was being followed by at least five or six starry-eyed males as she made her way. Some more confident suitors had hitched rides, presumably to secure their date for the night. Admirable enthusiasm! Often these happy couples

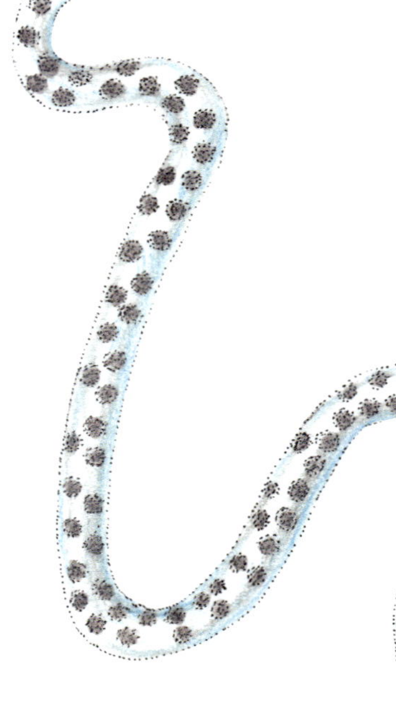


appeared to have gathered a small audience of unsuccessful suitors who looked on somewhat reproachfully, having lost the girl. And, as if to press home the point, the jockeys would offer up little squeaks of irritation, 'scoot it, losers!'

I've been seeing the occasional toad since late February, when perhaps the rare nights of slightly warmer temperatures have tempted them out of hibernation to forage for a midnight snack. Last night, however, was clearly the 'Beginning of Spring', as far as they were concerned.

I'm standing now at the far side of the pond, where it promises to be a little lighter as morning arrives. I've come to see if those toads made it this far.

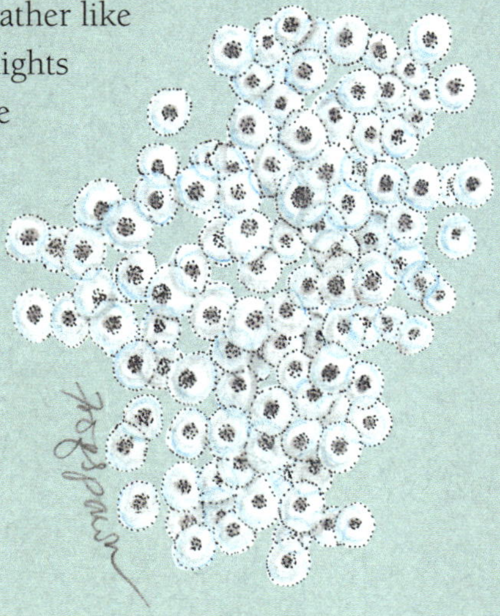

Billie's with me and she's dropped her favourite ball in the pond (thank you) in her eagerness to lap at the icy water. The steady radiance of ripples from her drinking spot makes the looking a little harder. Distracted by these silver-tipped waves as they slowly bloom, I allow my gaze to rest just beneath the surface, and now I can see movement. There's a toad… no, two toads. There seems to be a whole pile of them, each one partaking in a slow-motion performance of some free-form dance with no discernible tempo. For male toads, it seems that the drive to reproduce dissolves all sense of reason and personal safety. So keen are they to mate that they will hang on to fish, other males, even inanimate objects such as tree branches, in the water. Looking further along I spot a calmer, more sedate couple, the male atop and embracing the female, his forearms tucked neatly under her armpits as they very deliberately make their way along the shallow edge. Together they're unwinding a perfectly straight string of fertilized eggs, rather like you might unreel the Christmas lights before arranging them around the tree. I can also pick out several gently swaying clouds of frogspawn nearby; perhaps there's a safety in numbers when clumped that way. We've regular visitors – duck, moorhen and heron – all of whom are rather partial to toad- and frogspawn. And as spring progresses, dragonfly nymphs and water boatmen will also enjoy more than a nibble.

Walking along the edge of the pond, which has been transformed into a series of dot-dash roads, both major and minor, often intersecting one another, it's hard not to marvel at the very disciplined approach these toads have when organizing their nursery. If the eggs can hatch and make it through to become fully formed toads, they have one advantage over frogs in that their skin exudes a foul-tasting toxin when they feel threatened. This can help deter predators. Sadly, frogs have little in the way of defence, other than to hop away or emit a high-pitched scream.

Toads have enjoyed and endured a wildly contradictory reputation, at turns being seen as a portent of good fortune and as a harbinger of evil. As early at the twelfth century it was considered lucky to see one outside, but only if it crossed your path from the right (let's hope the pond was on the left then). However, by the eighteenth century if you found one in your house it signified that an enemy or great misfortune would come calling. It could also be a witch in disguise.

Nowadays most people regard toads with a level of indifference, which is a great shame since they are one of the best pest-control officers you could ever hope to have in your garden. Their diet is made up almost entirely of the creatures that dine on all our veg-plot efforts. With their habitats shrinking and their movements stymied by the introduction of new roads and developments, they have a hard enough time simply growing from tadpole to fully reproductive adult. However, their toughest challenges are those that they encounter on their journey to their old breeding ground.

This pond is perhaps forty years old. Toads can live up to ten years in the wild, and they tend to return each year to the place where they were spawned, so there's every chance that some of these intrepid amphibians might be veterans. I feel like cheering them on. But instead, I pick up a small broken branch from one of the old cherry trees and reach out to herd and pluck out Billie's ball from among these frisky creatures.

If all goes well, and these train tracks and clouds of spawn manage to reach full maturity, it's looking quite promising for their collaborative 'Frog (and Toad) Chorus', planned for later this year.

# Almost

Winter may yet have one last stab at disrupting the plans of spring but, by all accounts, we're on the verge, and I'm hoping there's very little it can do to stymie her progress.

I usually wake early but, for the last week or so, I've noticed the sun has beaten me to it. Having climbed the steep hill to the south of our little coombe, she's already casting her gaze across the valley and down to the town beyond, icing the neat regiments of roofs with the unique glow that only a new dawn can bestow.

At times I could curse my out-of-whack body clock, keener than I to rise, regardless of how inadequate the sleep may have been. But with a new season on the cusp, these early mornings seem like a gift. The incremental inching from late winter to early spring feels tantalizingly real, and hopeful. And with only a week or so to go before the vernal equinox, I've returned to my once-familiar activity of getting up and heading out. At our old home, in Sussex, I'd grab my dressing gown and boots and slip out of the house to take a walk through the woods and into our field, where I could stand awhile and simply be. It became my thing, my way of just taking stock of the previous day and its knotted string of conversations, moments enjoyed, and sometimes moments of stress. And with a little practice I found that I could unravel some of those stress-related tangles, which would free me up to look at the new day

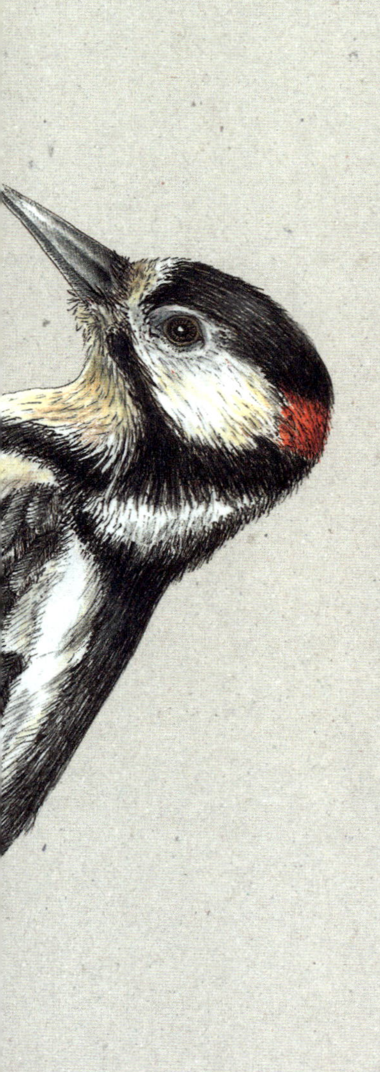

with fresh eyes. I cherished those minutes, sometimes even an hour or two, before daybreak. I often didn't have to share them with anyone else, since most of my family were still deep among the folds of sleep. I would take note of any tiny changes in the field or forest, look for signs of wild visitors and listen out for the first members of the morning choir as they warmed up for their grand performance each day.

This morning I was woken by two male blackbirds. Each day, regardless of the weather, they set up camp in and around the large but relatively naked hydrangea bush to hurl abuse at one another. It started about a month ago, and I've noticed that as the days have grown longer, so have their spats. Today, their beautiful argument began at perhaps 4.20am and they both clearly believed they were justified in their grievance, even if it meant going over old ground. In fact, old ground is what it's all about. Territorial disputes become even more pressing, with each bird keen to establish his domain, as they settle in for the breeding season. It seems to be the same for the wrens and blue tits, who like to chime in a little later in the morning.

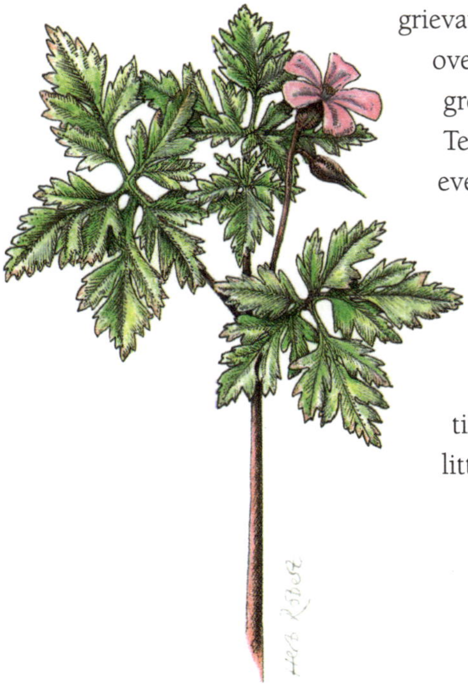

Recently, while kayaking along the estuary and out to sea, I've noticed that the same slanging matches have grown more audible. No longer is there the careful coyness of those trying to avoid a predatory gaze. And my late-afternoon shoreline wanders have become a feathery front line. All caution has been thrown to the onshore breeze and the air has become thick with the chatter of new couples organizing nesting arrangements. There's the sweet refrain of males puffing chests while beak-deep in dispute, while others are still engaged in the fiercely competitive game of 'Come Mate with Me'.

The weave of tiny lanes that loop their way around the villages and hamlets here are banked by deep, high hedges. In recent months, both farmers and the local council have worked hard to prepare them for the year, cutting back wayward branches, squaring off edges and topping them out to ensure clean visuals ahead for all that use these roads… all in an effort to avoid the need for maintenance later. As they race towards the growing season, these hedgerows will morph into thriving nurseries to be filled with this year's young.

The bare, weatherworn arms of beech and blackthorn that were so tightly woven as if to save heat and modesty are now beginning to fill with the first colour swatches of spring's new wardrobe. For the last few weeks, yellow primroses have been holding the stage door ajar for others to follow. The small, flushed faces of herb robert can be found peeking out from between early blades of bluebell foliage, and young nettles have appeared in clumps among the tangled skeletons of last year's brambles. For the lanes that wind closer to the coast, the hedgerows are beginning to change into a different spring collection. The sea air offers a damper, salty environment. However, a fierce storm can quickly reduce all lush growth to a severely abraded muddle. And yet there are those plants that seem to like nothing better than this harshest of habitats. Poking out from the ancient and rather dishevelled beards of traveller's joy are the lush leaves and yellow mops of young alexanders. Every exposed section of stone walling wears a carpet of dimple-leaved navelwort and, tucked into the darker crevices, small flashes of purple reveal the violet to be a hardier plant than I originally thought. Lesser celandine is beginning to flourish and spread like a spill of sunshine along most lanes, and especially those that snake towards the sea.

I hear from a friend in Sussex that the blackthorn is in bud, its tiny tight fists beginning to loosen. It was a sight I used to inwardly applaud when cycling along the lanes there. But here in Devon, we seem to have an abundance of wild cherry instead. Like torn clouds, fallen and swept into the corners of fields to be caught among the outstretched arms of overgrown thickets, it is everywhere I look… So much of it, that I've had to stop the car to hop out and double check that it really is cherry.

When out walking I'm now beginning to spy flooding tides of wild garlic. Great lakes of it appear to be leaking out from the more northerly facing, wooded areas around here. Until this revelation I'd been holding back on a jar of pickled buds from Sussex. But today I'll open them and enjoy their sweet crunch with a wedge of local cheese, happy in the knowledge that I'll soon be preserving more.

This year, more than any before, winter outstayed her welcome. Every sign of spring's approach has felt like a perceptible step towards hope, becoming the foundation upon which we might rebuild our lives.

Waiting for the kettle to boil, I open the kitchen window to be greeted by the resounding drum of a woodpecker.

Almost there now.

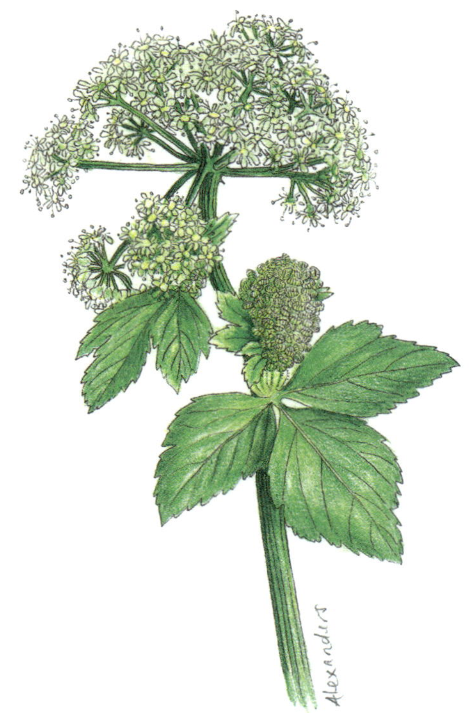

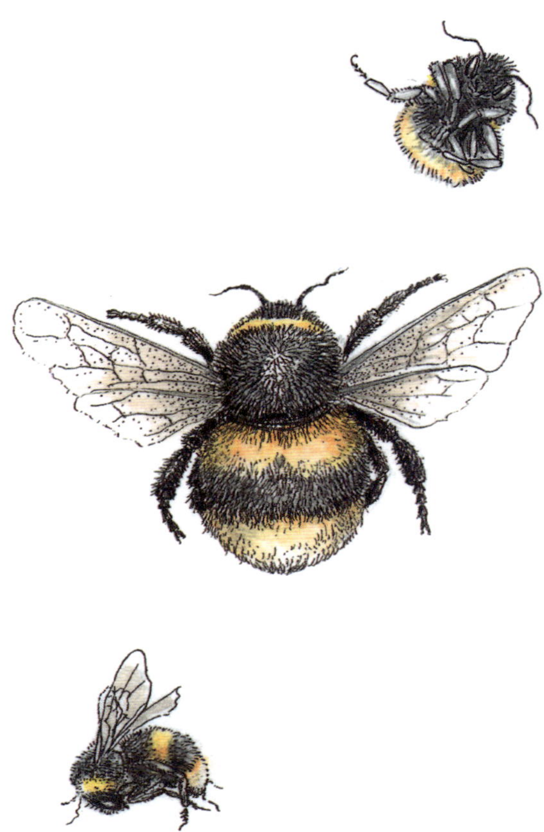

Buff-tailed Bumble Bee

*12 April*
*Morning lit, sky clear. Warmth on my back!*

*Standing near a patch of wood anemone, watching a buff-tailed bumblebee* (Bombus terrestris) *queen in search of the perfect nursery.*

*So many of them now, thundering around, lazy looping among the laden arms of the cherry tree; dropping to knee height for a closer inspection, then finally settling to rootle among last year's leaves and this year's blossom.*

*They've a particular fondness for an empty mouse nest.*

*Once she's found the ideal hole, she'll get to work, taking wax secreted from her abdomen to create a nursery for her colony, where she'll lay the first eggs to begin her family. Some of her firstborn will hatch and head out to gather nectar and pollen, while others will stay within the nursery to help nurture the young brood, maintain the hive and care for their queen.*

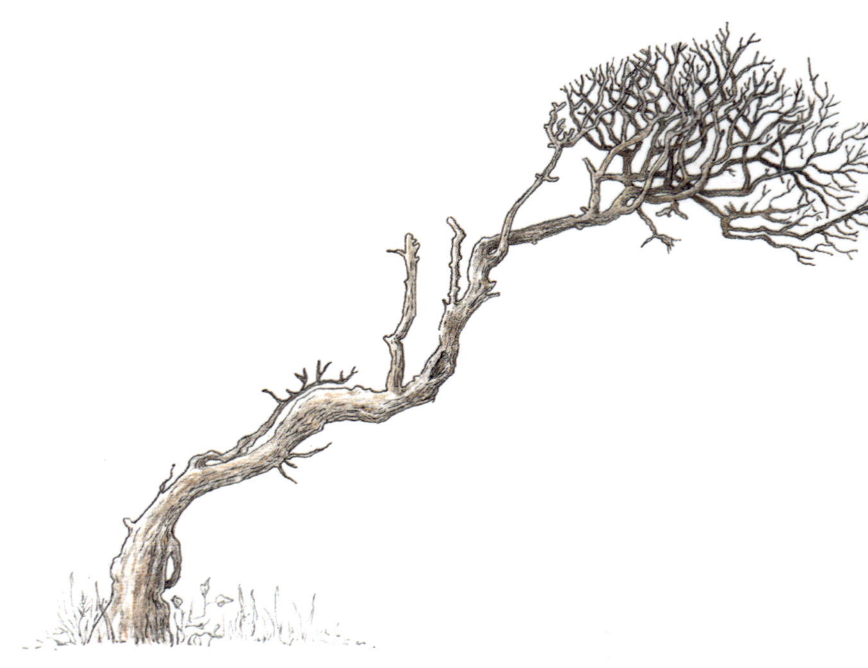

# Coast to Coast

I'm heading away from the soft rolling green of South Devon this morning and driving up to the north coast of Cornwall. I'm due to deliver an oil on canvas to a friend there. It's already feeling less like work, and more like a mini break, because I haven't driven this far since we moved last year.

Both windows are down, and the painting and I have just crossed over the River Tamar into Saltash; the sun has graced us with a smile broad enough to daub fresh freckles on my arm that's out, fingers splayed and riding the flow of Cornish air.

I don't know why I should be surprised any longer. Every time I step foot into this county it's the same. There's a slow and definite sense of being immersed into something otherworldly… After just a day of being here you leave feeling refreshed, exhausted and ever so slightly and delightfully unhinged.

When I turned eight, my parents made the decision to move us all to North Cornwall. It'd been something they'd been planning for a few years. I had no idea. My brother and I only knew Cornwall as 'the place that takes eight hours to drive to, and seemingly (sadly) no time at all to leave'. We'd holidayed there from a very early age, always trailing to the same cup and curve of coastline… bucket and spade in hand, and a promise of Princes-sardine-and-tomato-paste sandwiches. So, to us, moving here seemed like a jackpot win: the guarantee of a never-ending holiday.

The reality was, of course, a little different. As idyllic as postcards may be, we all found out that living in Cornwall is not the same as holidaying there. As money became tight and then tighter still, life became more than the adventure that my parents had planned for. Clothes were all second hand, mended and extended. The 50p slot-meter for the electrics was reviled and sworn at as much as the bank manager. Shampoo du jour was Co-op's own-brand washing-up liquid, and dustbins were lined with newspaper, not to cut down on plastic waste but simply because we couldn't afford bin liners.

Mum and Dad grew, caught, made and bartered for most of what we ate. Along with a fair amount of fish, we ate rabbit, pigeon, pheasant and liver (tubes and all, 'just cut them out, darling…'). I'm certain there was a huge variety of food, but these ingredients stick in the mind more than others. Also rice pudding, and cooking apples every which way possible. And every meal was genuinely delicious, even the liver.

There was no central heating but we had our little Rayburn, which performed the daily miracle of warming, drying and cooking all that was needed to keep tummies full, and sanity in balance.

School was a bit of a struggle. Arriving from another part of England, I was considered 'posh' and so it was assumed that I must live in a posh house, eat posh food, while wearing posh clothes and discussing posh pursuits. Our parents rarely took us to play dates or the beach, since Dad was always working and Mum didn't drive. Much of my free time was spent outside, walking along the edges of streams, counting badger setts, climbing overgrown field boundaries to explore and chatting to imaginary friends who were a little more forgiving than some of those I'd failed to charm at school. And so, to a certain extent, my brother and I became landlocked in this little pocket near the coast.

We had our chores and beyond that it was a team effort, whenever required… chopping wood, mucking out chickens, cutting stingers, gutting fish or bleaching the mould-mottled windowsills of the cottage in which we lived. We just got on with it.

The strangest thing is that, although there were many pretty appalling scrapes and hard times, I don't remember ever feeling that my life was 'tough' or untenable or feeling envious of the seemingly bright and spangly lives of friends and visiting cousins. I'm not sure how they did it, but my parents never betrayed any sense of the futility and frustration that, looking back, they must surely have felt at times. I may have been that miserable teenager but I never once considered that my life was lacking. It's not as if I didn't know how rubbish things were. I did. It's just that they never overdramatized it. They just got on and did their best to rise to each challenge and move on. And so we did the same.

As an adult now, I can see why their enthusiasm never waned. If you're not vigilant, the chapped and chafed edges of the north Cornish coast tend to creep under the skin, to work their way beneath the weave of fascia and beyond. And, if you linger long enough, they will slip between the fortress of ribs and find their way to your heart.

The criss-cross spread of skinny lanes that lead you to the weather-beaten beaches hold a charm of their own, particularly in the warmer months. Wind-walloped hedges get dressed for summer: honeysuckle, vetch, hogweed and sheep's-bit, crowds of cow parsley and clusters of yarrow. And as the weeks continue the colours become more vivid, with flames of wild crocosmia and walls of pink and red valerian. Most are crowned with bushes and trees that grow stooped but sturdy, bodies contorted in their efforts to thrive. Early spears of babington's leeks launch out from the top, while looped and leggy flowering bramble will grab at an arm as you drive by.

And then there's the sea. The Celtic Sea has a different bite to the English Channel. South Devon waters will usually lull and lap to tempt you in. Of course, there are storms; but they are slow to anger and soon blow over, brows unfurrowing, to offer a calm passage to all who enter. The Celtic Sea, however, can be mercurial; her temper can become tempestuous with very little goading. She will throw herself into a raging strop that can gather up and crush the unsuspecting and innocent, often leaving them broken, sometimes lifeless too. Only later, when she's quietened to simmering, does she appear contrite and full of remorse.

Whether a gentle caress or a searing slap, the Celtic Sea is at once enthralling for all those who choose to float on her, swim in her or paddle along the edges of her reach. Everything about her is utterly bewitching. For me, my state of joy grows in proportion to the increased knottage of wind and foul weather, such is my somewhat perverse definition of 'wonderful'.

My parents gave me this outlook, a touchstone to help discern what is good, what genuinely matters. There are often times when, as an adult, I've had some fairly tricky challenges to wade through, days when not a lot seems to be particularly good, or indeed wonderful. It's the same for many of us, isn't it? And unsurprisingly, this has never been more relevant than during these times of the pandemic. However, this yardstick that I have has proven to be a saviour of sanity and humour. This and, of course, Cornwall.

Painting delivered, I can't resist a quick detour to one of my favourite beaches before driving back home. I pull into a tiny car park, car bonnet nosing a sprawl of wild fennel. I hop out and hurry along the lane to the narrow slipway that opens on to this diminutive bay of rock and stone. To many it looks quite inhospitable – there's no stretch of perfect sand, no public loos or an ice-cream van. It's rare that people stay for long. If they did, they'd know that when the tide rolls out it reveals the softest carpet of

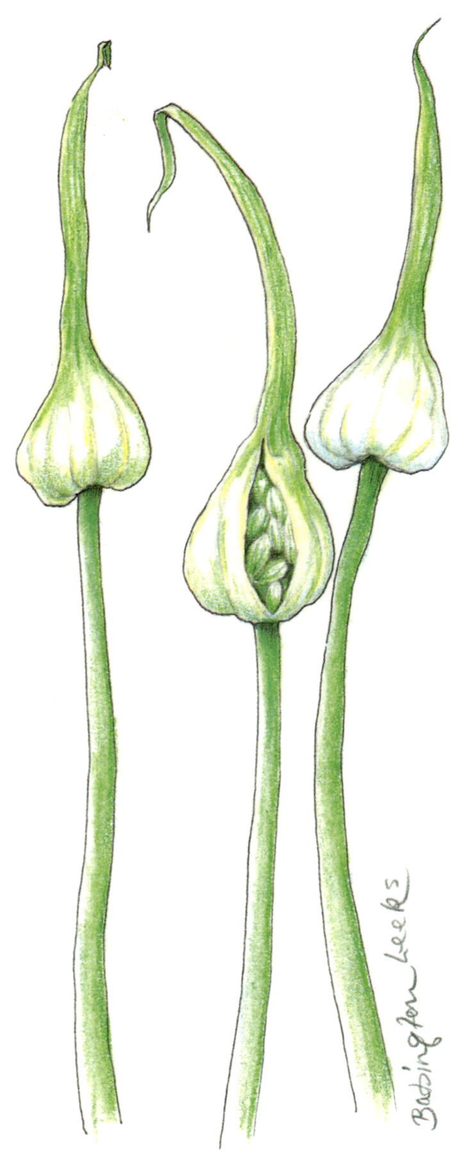
Babington Leeks

sand, and that you can paddle out to walk among shifting clouds of silvered fish. However, most holidaymakers are impatient for the picture-perfect beach in which to set up their windbreak, turn on their radio and snooze while their children dig moats and mould castles. Today, the tide is in, the sand is hidden and the rock pools swallowed whole. There's no one else around.

It's here that I have my first memories of drawing. I used to pick up sea-smoothed tablets of slate, rootle around for a shard of broken limpet, and then sit among the stones to inscribe my slab with lines and curves. Faces, feet, multi-petalled flowers and fantastical fishes: an odd mixture, but those were the things I wanted to portray then. Comfort would settle like a blanket whenever I did this. I used to sketch on stones a lot.

I'd very much like to stay a while and draw, but it's time to head south again.

Home now, and the sun is still owning the sky today. I'll grab a mug of tea, then head to the beach for a swim.

It's only been a little over nine months, but I sense that the gentler, tree-lined coast of South Devon is no less devious in its campaign to ensnare me.

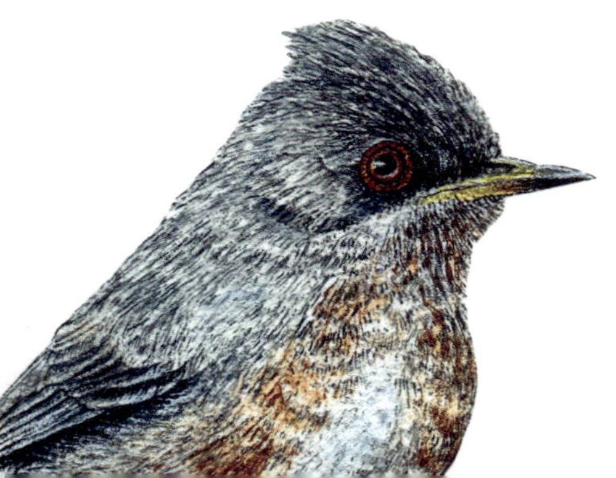

# Here

I think that because of the enduring fascination I have with our new surroundings, I've found it impossible to sit still. I have a creeping sense that, even after a relatively short passage of time, I'm starting to become enveloped in the very fibre of this coast and, as a result, the final shudders of spring as she throws off winter's chilling grip are almost palpable.

I can feel it in the way the coldness of the sea no longer bites so deeply. When I emerge from a swim, my skin isn't as darkened with the surge of blood to the extremities... rosier now than ruddy-to-purple. It'll be a while yet before the sea reaches for the double digits and me for the lilo, but we're moving in the right direction. I don't think it can get any colder now.

A ripping tide will still try to turn the nose of my kayak as I paddle along the river and out to sea. However, the colour of the Dart no longer mirrors a thunderous sky and has brightened to reflect the longer, better-lit days. And when the sun manages to push through a caravan of clouds, her increasingly higher arc from east to west enables her to reach down and trim the ends of winter's long dark shadows.

Recent hikes have found me rambling along the tree-lined clifftop paths, yanking down my coat zip, removing gloves and wishing I'd worn shorts rather than jeans. But

I'm not daft enough to think we've seen the last of the snow. It is only April, after all.

Indeed, yesterday, after a spate of grey-flanked but warmer days, the wind shifted direction and a chilled northerly sauntered in. Even though it was relatively gentle, there was an underlying steely determination in its demeanour and, within hours, it had managed to persuade the loitering cloud base to move on from the coast. So, this morning we woke to azure skies and a sun so bright that you could have been fooled into thinking that summer had walked through the door; until we opened it and stepped out into the sharp cold.

Donning a coat, scarf and anything else that might plug the gaps against the wind, I packed a small bag with essentials (water, phone, ginger biscuits) and together with a friend drove to the coast.

It'd been deep winter when I last walked this path and it was almost unrecognizable now that spring was in charge. We wound along the trail lined with budded beech and oak, paused but ready to explode into leaf, while numerous young sycamore whips were already waving small green hands in the wood-buffered breeze.

With roots dug deep to fasten them to the steep cliffs, the older, weather-worn trees here grow in whispering distance from one another, perhaps safety in numbers being their objective against this rather inhospitable choice of home. Some of their roots rise to crest the earthen path – snake-like and polished by years of human footfall – before dipping down once more to grasp hold of a neighbouring root or a layer of rock.

We paused to take in the scene. Where the ground was once mostly bare – save for the usual muddle of old brambles – a torrent of bluebell shoots and acid-green dog's mercury had filled in every cranny and crevice so that the trees appeared to float atop this lush, sloping overlay of green.

With their flowers mostly gone, a generous sprawl of winter heliotrope carpeted the verges of our path through the woodland. As the track opened out on to the edge of the cliff it was held in check by clotted-cream clouds of blackthorn. So much of it! Just when I'd assumed that I'd not find any this year. We kept to the edges of the land, the heavily battered ramparts of the slate cliff below peaking up through the bank of sloe, when our range of view suddenly widened, and the path steepened. We climbed up, turning inland a little and made our way to an open expanse of heathland punctuated with low, sunny thickets of gorse. Here the wind found us at last, as it slipped down the hill towards the sea below. We turned up our collars and hitched up our shoulders against its prying fingers. But then I heard a familiar sound; borne on the wind's back was the sweet bubbling chatter of skylarks.

We stood awhile, listening, peering up the incline and into the wind to see the choristers. Not a chance. Eyes streaming, admitting defeat, we started walking again. The path curved once more, leading us back to the edge of the land and this time into a loose tangle of thorn, some wire fencing and a series of swerves through a bramble gauntlet. Once out the other side we were greeted by a crowded bank of violet-tipped sea cabbage.

Then, finally, a clear view of water.

With the sun on our faces we scrambled, hopped and slid down a partially collapsed ladder, a casualty of the storms this coast has endured over winter.

It's a strange thing that occurs when we have to work a little harder than originally expected… The sense of 'achievement' seems to embellish the denouement with a wonder and magic that surely wouldn't have been present if the reach had been easy.

This little cove held an air of enchantment. Shaped like a diminutive amphitheatre, it was banked by steps of grey-green slate that led the eye down to a slow-shifting tide of turquoise and teal. The steady breeze left our backs as we stepped over the last broken rung and on to sand at last. With very little thought we both stripped off entirely and ran into the sea. The scene switched almost immediately. A cloud the colour and shape of a wood pigeon's wing slipped across the stage and snow began to fall on our upturned faces. We whooped and laughed, marvelling at the sudden change. Swimming out to the edge of the cove we stayed and swayed a while, treading water and admiring the new view of the bay. The sea was slow to nip, but our fingers and toes ached eventually, and we reluctantly swam back to the beach. With no towels to dry us we hoped for

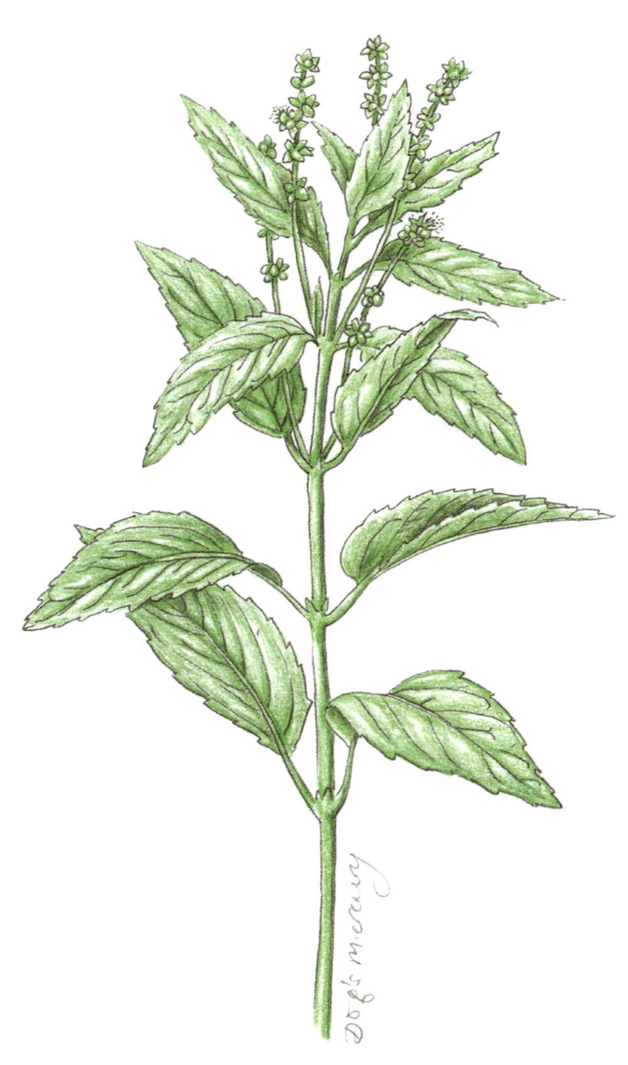
Dog's mercury

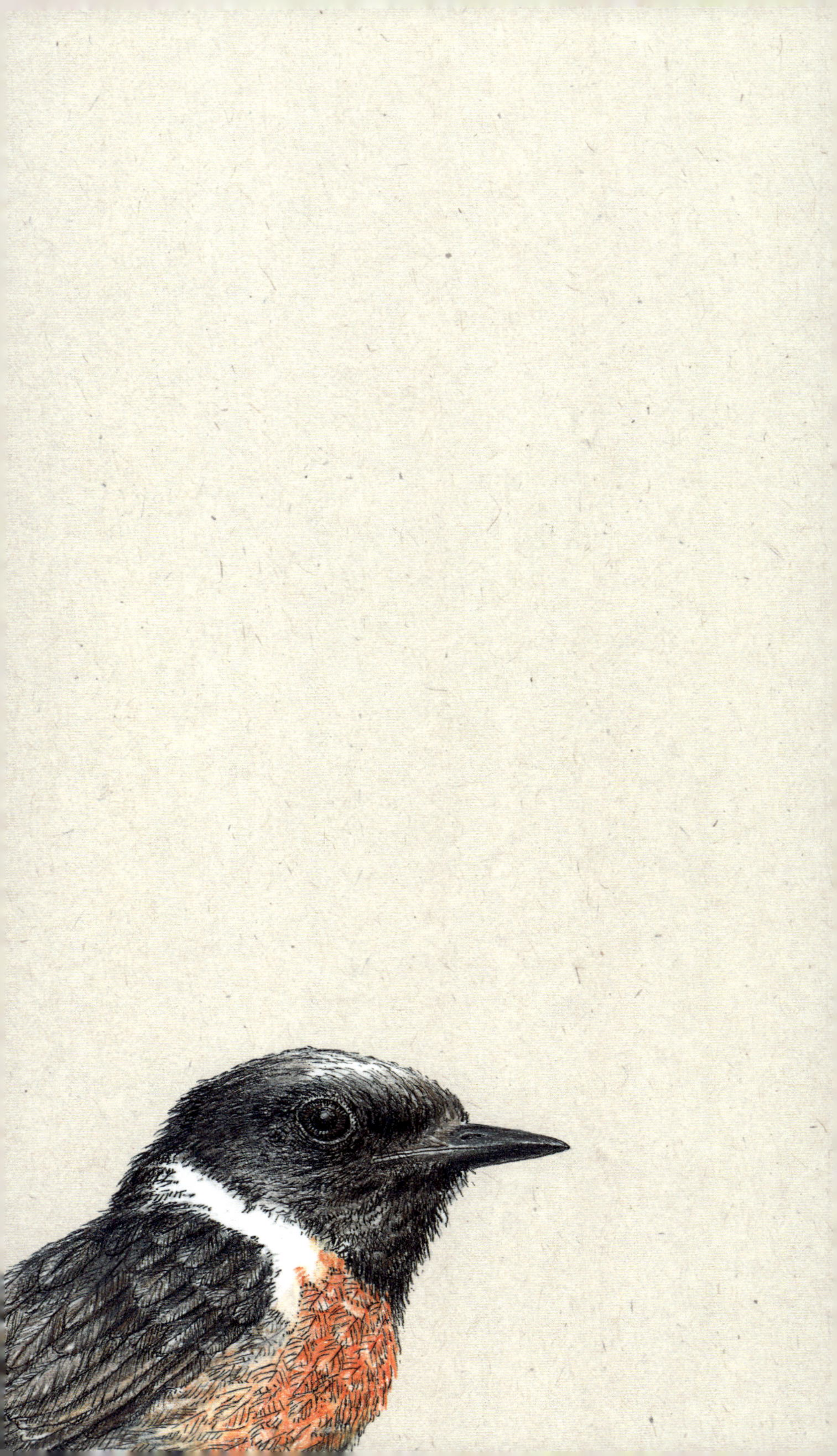

a return of sunshine if only to buffer the rising tide of cold. As if on cue, the unravelling cloud left the scene and the sun stepped forward once more to light the sea and warm our limbs. Inching wet bodies back into snow-dampened clothes, a fine tilth of sand to add to the joy, we wiggled toes into socks and pulled on boots before making our way back up the broken staircase.

Back out on the bluff. And we found a new gathering of choristers had taken over from the skylarks. I wish I'd brought binoculars to see more clearly this swooping quartet of plump and cheerful souls. Their presence here, demeanour, call and behaviour seemed to tick most of the boxes for the Dartford warbler. But for all my enthusiasm, and rather rubbish eyesight, they could just as easily have been a very chatty club of dunnocks on a family outing!

I'm back this afternoon, now with binoculars. So are the birds. At last, I can see their markings and they're not Dartford warblers but a handsome family of stonechats... male, female and juveniles too. With no wind present today, their call is clearer and makes me laugh. Their sharp chirrup and chatter really do sound like two stones striking against one another!

I'm hurrying home now to grab my pencils.

*11 May*

*At least half a dozen red-tailed bumblebees filling up among the wild garlic flowers. A first for me. Such an orange glow to their rears!*

*Also, a bee-fly, long proboscis dipping in to feed on the nectar of the last primroses. Like little hummingbirds. Less than desirable behaviour when it comes to parenting! They are by nature parasitic, depositing* their eggs in the nests of solitary bees, beetles and wasps. Once hatched the young will feed on the eggs and larvae of their host. Not such great news for the owner of that nursery.*

**They have a special pouch under their tail to store sand in, and they roll their eggs in this to add a little weight to hurl the eggs successfully at the chosen nest or burrow. Enviable shot-put skills.*

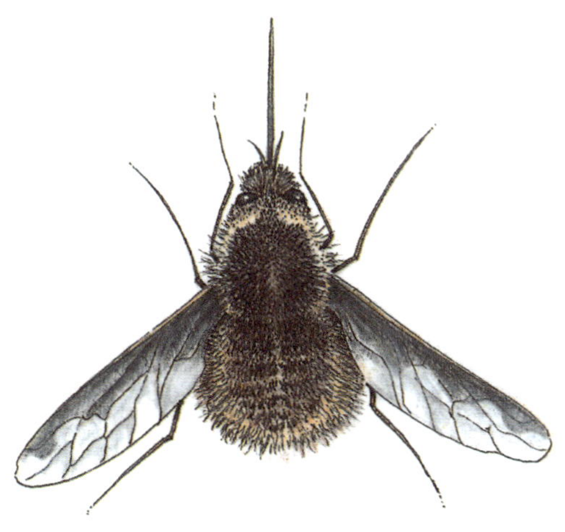

*Bombylius major*

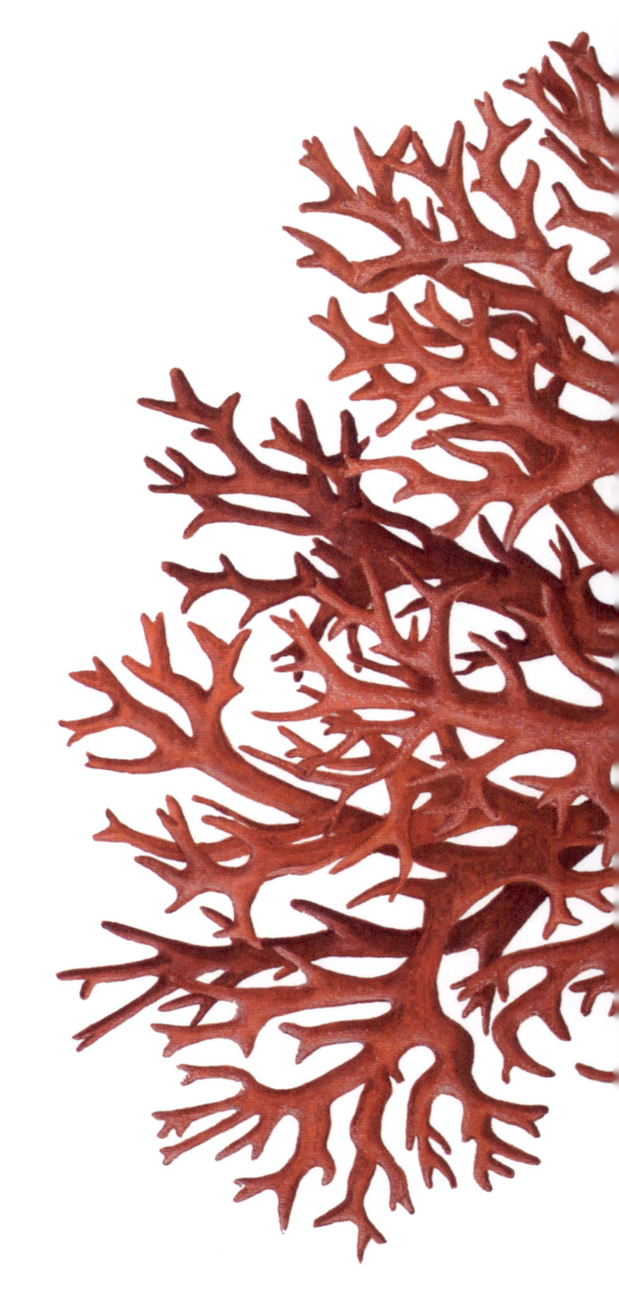

# When I Grow Up

I've been swimming in and rootling along the edges of the sea since the day we moved here. And I've come to realize with more certainty than ever just how dependent I am on her. It's inevitable. She has that effect on many, I think. I've read and enjoyed many romantic tributes to the sea and written of her in similar (though possibly less eloquent) prose. But perhaps my perspective could be seen as rather self-centred and self-serving. It may go some way to swelling a reader's empathy towards her but does it move them to do something to support her, as she undoubtedly supports us?

We got our first colour TV in 1975, when I was seven, upgrading from a black-and-white Philco screen no bigger than a pocket handkerchief to one of full glorious, tuneable colour. You could adjust how much oomph you wanted the picture to have. And, of course, I wanted it at full impact, with the occasional flare of fuchsia when I turned the knob a bit too far. Fabulous. It was possibly a little bit too glorious, bordering on gaudy for my parents, but I was absolutely having it all.

After-school TV became a real thing. Favourites were *Blue Peter; Animal Magic*, with Johnny Morris lending his gently comedic commentary to every elephant, monkey and parrot; the very wonderful *Take Hart*. There were many more shows, of course, but it was Saturday mornings that held a special place in my heart... That was when I could dip into the wild and exotic water world of the Frenchman, Jacques Cousteau. As an inventor, naval pilot, filmmaker and conservationist, he could also claim the title of 'Most Charming Environmental Disruptor' of his time. With his extraordinary zest for underwater research and discovery and his undeniable ability to use the evolving mediums of film and photography, he managed to take a whole generation of armchair enthusiasts with him beneath the water to see this thriving garden full of such wonder and vibrancy. We went to Corsica to seek out the last of the red coral, we dived at night to scout for the fantastic living fossil, the nautilus, and we listened to humpback whales sing as they migrated. These films evoked a sense of complete awe in me; that such places and animals existed seemed almost fantastical. However, as each episode ended it also left me a little deflated and wistful, since I was almost certain that to visit these extraordinary places and creatures for real was unlikely. This wasn't only because I was eight and couldn't book a flight/boat to go and join Monsieur Cousteau but because, throughout his films, it became very clear, even to a child, that this underwater world might become diminished and reduced to little more than a pale skeleton by the time I grew up and had amassed enough money to head out and see it for myself.

With my father being a furniture restorer, we regularly had customers turning up at our house to collect their beloved sewing table or ancient armchair. I used to hang around to spy on them, and occasionally get found out and pulled in to make small talk while Dad helped to wrap and pack the furniture into their car. It's a curious thing that some adults, when meeting a child for the first time, will almost invariably feel compelled to ask one question:

'So, what do you want to be when you grow up?'

It's a tough one, for some children can let their imagination run riot and spontaneously let rip a fabulous fountain of possibilities. But for others I suspect there is nothing more paralyzing than to be pinned down by such an inquiry. Good god, I wanted to be everything and anything, and it could flip instantly, depending on day, mood and recent TV viewing. The brave child who excitedly pronounced their chosen job title would from then on be reminded of their career choice every time they saw that grown-up again.

My problem was that I thought my dream choice of career would seem stupid to a grown-up. It certainly didn't fit with the reality of our lifestyle and budget. Instead, I decided to keep my mouth shut and would always say that I didn't know.

Deep down, however, among the multitude of dreams I quietly nursed, there was one that I'd return to time and again. I didn't know the name for it then; I'm not even sure it existed as a job. But, like Cousteau, I wanted to go on expeditions to far-flung oceans to study their flora and fauna, and ultimately to illustrate it for those who might never get the chance to see it.

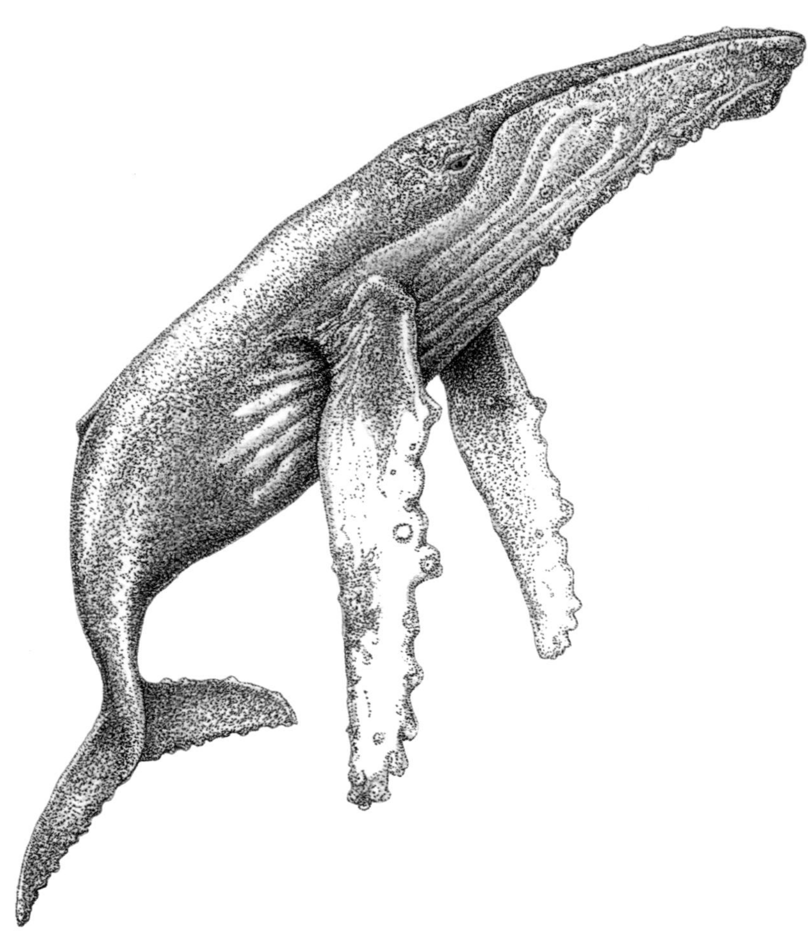

However, with A-level choices looming and advice from a rather bored careers officer that I should be something sensible (a physiotherapist or – perhaps because of my accent – a speech therapist), I lost that sense of possibility and excitement, and floundered.

The circuitous journey to finally picking up a pencil or a brush again took a decade or more. But the enthusiasm was never quite extinguished. And so here I am, an illustrator. Perhaps if I had been a bit more ballsy to state my dream job when those kindly but awkward adults had asked 'that' question, then I might have arrived earlier. But I'm here. I'm at liberty to research, study and illustrate the natural world, and I do feel incredibly fortunate to be able to call it a job. Whether what I do makes a difference to world ecology or the urgent challenges facing our environment is debatable at best, and very doubtful.

However, I feel connected again. Being here on the coast once more has allowed that association to grow roots and I have rediscovered the fierce loyalty I had to the sea. So, until I can think of a more impactful way to support her, I am that woman who wanders along the shoreline, gathering up plastic, metal, torn pieces of clothing and fishing wires. I'm also that woman who chases after driving litterers and pops their discarded non-recyclable coffee cup back through the open window of their car when they become stuck in a traffic. Sorry if that was you. But seriously, take it home!

# Summer

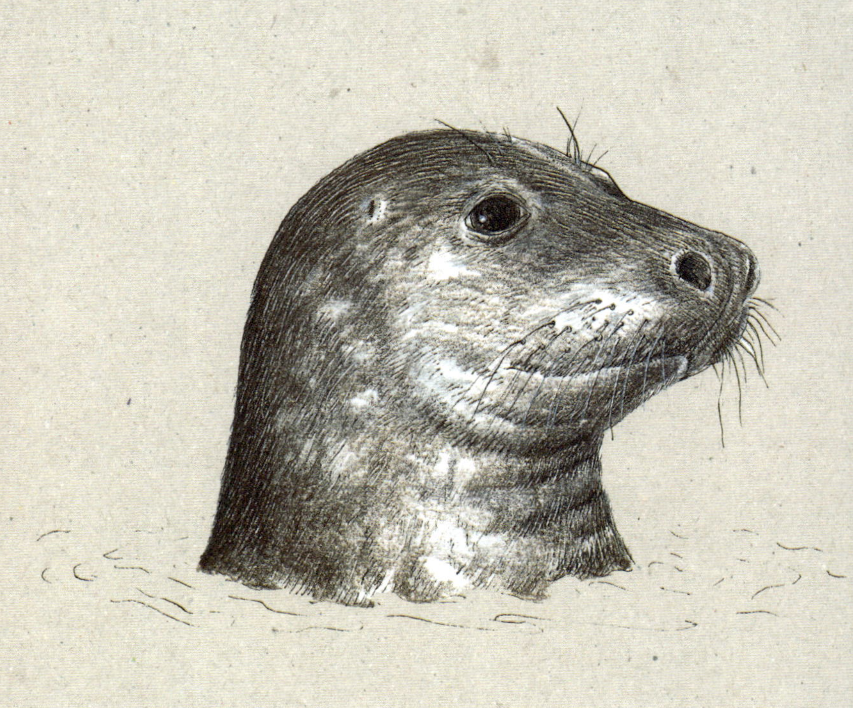

# Their Element

Back in November, having woken to another sky the colour of dirty slate, we decided that instead of taking the easy option of lingering over a pot of coffee, it was time to get to know the broad and serpentine ways of the River Dart, which very much defines this stretch of the South Devon coast.

Piling on layers of clothing, gloves and woolly hats, we squeezed into our kayaks, flipped over the drum-tight rims of our 'skirts' and launched from a pebbled beach. We nudged our way through the murk and out into the belly of the Dart. With the river in lull before it began to rise, we made good progress through the slack waters, marvelling at the perfectly trimmed undercarriages of all the low-lying oak trees that have been shaped by the ebb and flow of the semi-salty tide. These native oaks are thought to be the reason why the river is named the Dart. The word 'dart' is said to stem from the Celtic word for oak, and this makes sense, since as we paddled along it became apparent that these slightly squat and stocky trees cover almost the entirety of both banks of this wide, tidal waterway. The yawning, mud-flanked route, running through the land and out to sea, was carved out over millennia by the bare hands of nature. And yet, even in their leafless state, the crowded shoreline of oak lent a sense of elegant grandeur more reminiscent of royal parklands.

We paddled and chatted, enjoying the cold air that tugged on tear ducts and noses alike. Cheeks ruddy, and bodies swaddled, we began to morph into internal combustion engines as we steered our way through the dark waters. We counted cormorants, laughed at the barking mallards, and marvelled at a little egret; a white slip of a thing, stooped and perfectly mirrored in the glint of the steely shallows. Hats and gloves were removed and tucked into lifejacket pockets as we began to quietly poach with the continued effort of paddling.

But then we saw the grey seals. I only became aware of their presence when I heard a gentle snort of water from my port side. I turned to find two dark-pebble eyes studying me. They were spaced wide below a slightly domed, dark and dappled forehead; several brow hairs sprouted at rakish angles lending their owner a slightly cartoonish, surprised look. A long Roman snout ended with parallel nostrils that occasionally flared, and I was reminded of our dog, Billie, and the way she sniffs the air to assess a situation. Its muzzle was peppered with a generous covering of whiskers, its neck, comfortably thick with a suggestion of a double chin! My guess is that it was a male and, as he held my gaze for a while, he felt strangely familiar, friendly. Stupidly, I felt the overwhelming need to say something. So rather predictably I said, 'hello'. Several times, too. Ridiculous. He regarded me a while longer, no hint of disdain at my foolish efforts to do a Dolittle. Finally, he blinked, then soundlessly slipped beneath the water.

Stunned and inwardly squealing, I looked at Marc with no doubt the daftest of grins plastered across my face. He grinned back and we wondered whether our new friend might pop up again. With the tide still in flux before it began to creep back in, we knew we had limited time to reach our destination before we'd be forced to turn back. So, taking hold of our paddles, we began to push on.

Even before our kayaks birthed a new wake we heard a snort, and there he was, this time in front. I said 'hello' again, feeling just a little less idiotic, since he remained buoyant and watched as if to appraise our relatively clumsy, splashy progress. Then, as if to make it clear who owned this river, he raised himself up to reveal his light-coloured barrel chest, then plunged nose first. His slick and silvered body followed, his tail the last to disappear.

He sliced through the water much like the swallow through the air… effortlessly and unthinking. It occurred to me that both have full command of their element, leaving us to be the audience, and the occasional, graceless pretender.

He appeared several more times during that paddle. Each time he seemed to be waiting for us to catch up, and once we drew level, he'd disappear again. Eventually he either tired of this 'game' or had other plans, and we were left alone to continue our journey. But I made a promise to myself that I would come to see the seals again.

We're now edging towards summer. To my joy, swallows arrived last month, and they are settled and sitting. I occasionally spy the curious face of a female peering out

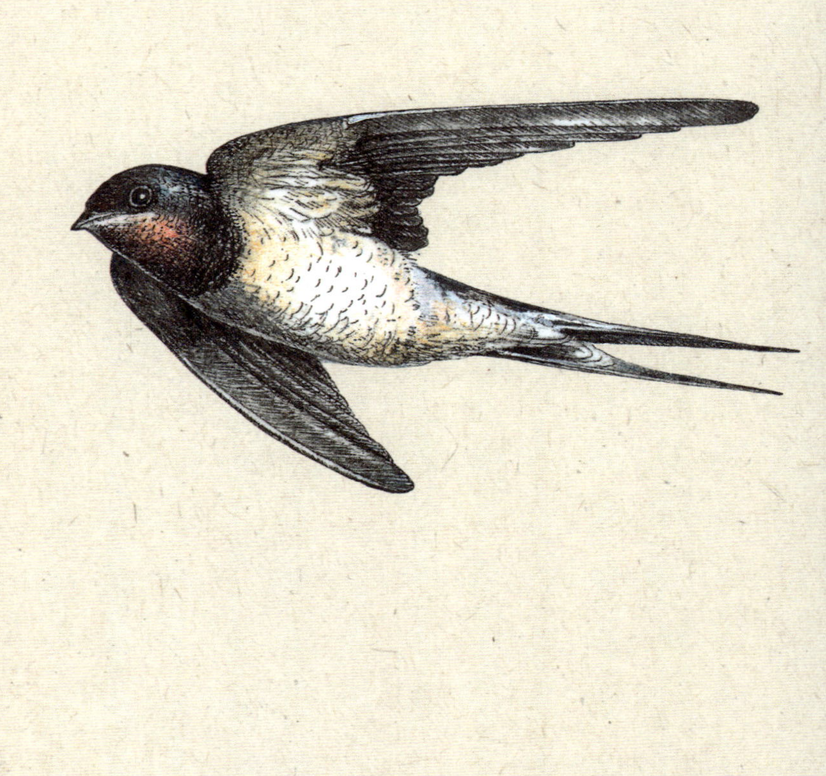

from among the eaves of the barn. They will incubate their clutch of eggs for up to fifteen minutes at a time, then swoop out to cruise the air for a snack before returning to the job in hand. I marvel at their self-discipline not to go out and spend the whole day feasting. Thinking of my own incubation periods (of months rather than days), I can wholly relate to this need for food and a change of scenery. And in these longer evenings it has become our preferred viewing to sit outside and watch them and their partners dine on the airborne insects.

I've not been out on the water for a while, but recently heard that one of the grey seals' favourite sunbathing spots is within paddling distance from here. So today we've decided to head out to see if we might find them.

It's late May, and the sea is a mirror; every rocky notch and inlet is perfectly reflected, a fantastical show of toothsome grins and sleeping giants as we slide our way to the mouth of the estuary. I recently came across charts that show the tidal movement of the water along this route and out into the bay. The charts show a stream of arrows with some wildly adventurous ones preferring to flow in the opposite direction. To me they appear as an indecisive wave of migratory birds. The tide is now high but three hours into the ebb, and it's gently coaxing our kayaks eastwards. I imagine us both floating atop a raft of these birdlike little arrows.

Mewstone is an uninhabitable islet, lying just off the coast of Dartmouth. From Google Earth it appears as some early, ill-formed cousin of the tardigrade, with a hatch of young rocks close by in the lee of its westernmost flanks. But at sea level it is an up-thrusting fortress of basalt slabs. From a distance it's clear why she's called the Mewstone: on her shores there's a rowdy colony of common gulls, otherwise known as seamews. Some are quietly preening but the majority are either shouting or deep in conversation. We navigate our way a little nearer to some of the smaller, squatter rocks, the gentle lilt of water catching us to draw us close then pushing us away.

And there are the seals, at least eleven of them, all within close proximity of one another. They appear almost molten, fluid in form, as they lie draped across this, the sharpest of beds. A few slowly slide their gaze in our direction but most don't seem too interested. Then a large male the colour of brindled granite looks our way, his gaze almost stern. He opens his mouth, and it keeps opening; there are his teeth with canines proud. It's at such odds with the gentle, almost dog-like expression I've long associated with grey seals, that I'm thrown. This no longer resembles the sweet, playful creature portrayed in 'Save the Seals' marketing literature. This is a wild creature, wired for survival. Even from this distance I can feel my sense of wonder and delight becoming a little dampened... a renewed respect edged with a chill whiff of fear. I've read enough to know that if seals feel threatened, they will usually move away, but not always. Well-meaning folk have occasionally pushed the seal's limit of comfortable proximity, resulting in some unsettling reports of seal attacks. Though admittedly a lot of those reports were during their breeding time, not now in the summer.

The wide maw tilts back a little further and his eyes become slits. It turns out he's yawning! But I've no intention of paddling closer. We stay and sway a while longer, then with one last look, we turn to go. We've been out here a good hour or more and the fleet of arrows will have miraculously flipped. With any luck they will bear us back in the direction of Dartmouth, and home.

There will always be a natural curiosity of the human for the wild, and from what I've found so far, it's not always a one-way flow. However, the lines between respective territories can often be too vague and misread. We frequently overstep the boundary lines. And whereas we have options to go elsewhere for our adventures, they don't. They're not on holiday; this is their home.

*2 June*

*May bug, cockchafer, billy witch* (Melolontha melolontha)*… found in the kitchen sink.*

*A handsome male… seven little feathers to his antler-like antennae (females have six).*

*They always look to me like they've been put together by a mad Victorian inventor, such a wonderful collection of parts, some of which seem fantastical if a little mismatched. No surprise that their flying style is a tad clumsy and misguided, rather like a wind-up toy with a spinning compass!*

*Carefully relocated him to the fresh young leaves of an old oak, where hopefully he'll settle down for a snack.*

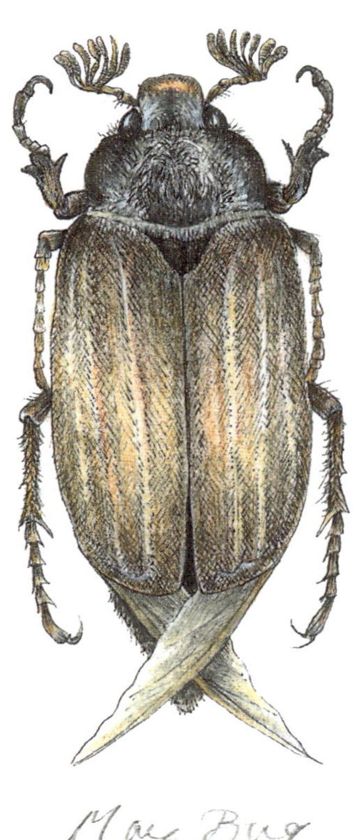

May Bug

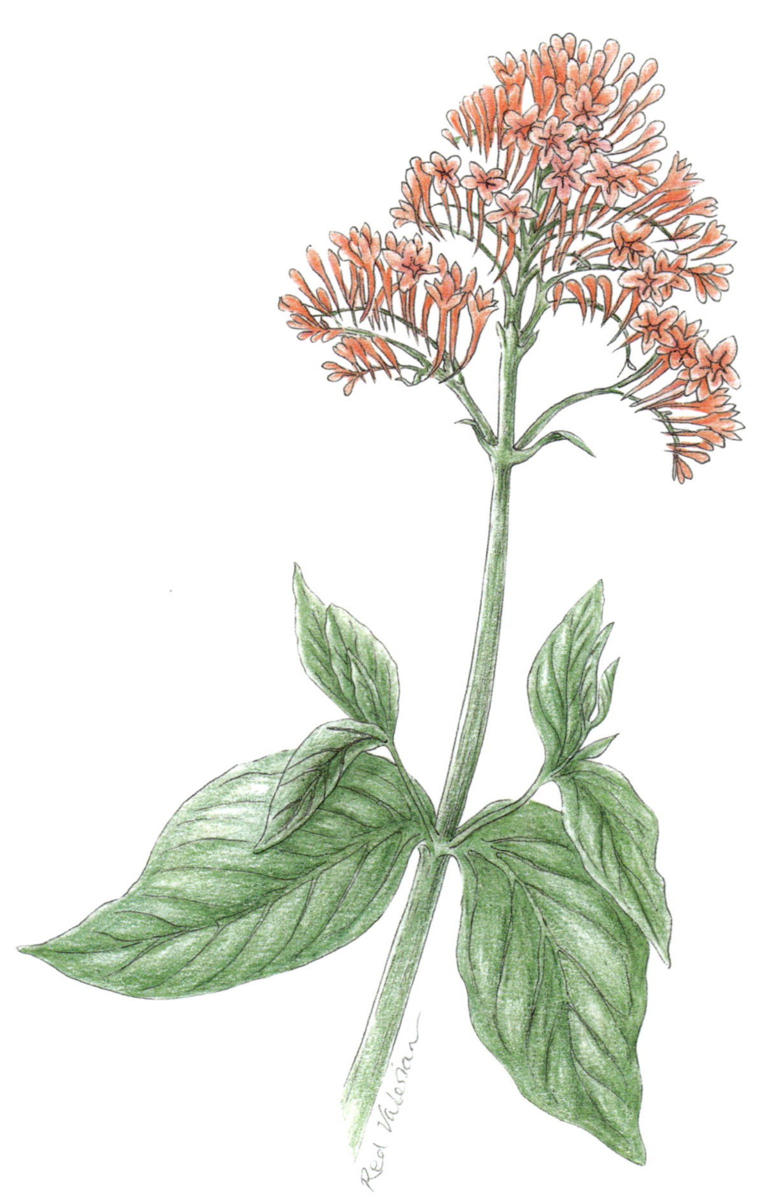
Red Valerian

# Red Valerian

Mid-June and the hedgerows are awash with dense clusters of red valerian. It's as if someone has raided the lipstick counter and decided to daub the high banks with random splashes of the most sensuous of reds and hot, hot pinks (perhaps that's why the flower is also known as kiss-me-quick). Every car journey I make now, I have the wing mirrors tucked in to compensate for the increasing bulk of the hedge. Whereas in winter and early spring their profile was slim and minimalist in palette, they now sport full, floral bellies that shiver and shimmy with every passing car. Of course, all hedgerows (if left untrimmed after nesting begins) should be wearing a brightly coloured costume of abundant wildness at this time, but red valerian is the one bloom that stands out along the lanes that weave their way to and from the coast.

I haven't seen this much since I left Cornwall at the age of eighteen. I was in such a hurry to stretch out beyond its borders that at the time any awareness or indeed affection I might have had for my home was entirely swamped by a rising tide of certainty that I was missing out on life.

But now, seeing the large and cheerful sprawls of this plant in flower is a joy. And it's drawing from a well of nostalgia for summers there, along her north coast.

Every day of a summer holiday would be spent going to the beach. It didn't much matter which one, because if we were all together it was the right beach. In truth there were only a few that my cousins and my brother and I would alternate between, each one being perfect for either surfing, swimming or sunbathing. Because of their locations we needed a parent to drive us to whichever one was deemed appropriate for the tide and the weather that day.

As we transitioned into teenagers, different beaches went in and out of favour. There was one that became an obsession; its width, scale and orientation along the coast rendered it the perfect beach for bodyboarding. So, with enough money to buy a pasty and rent a wooden board, we'd be dropped off and left to get on with being kids.

As we got older, my brother and some of my cousins managed to pass their driving tests, which meant a new game evolved to see how many people we could cram into a car. The car was usually of diminutive proportions, but to assume that this would lower the body count within would be foolish. The average tin of sardines will contain perhaps seven headless beauties. We were ambitious! We worked out that with careful planning and stacking, my brother's Mini could fit nine of us. This isn't something I've told our (now driving) children (or my parents come to that), but at the time we were ridiculously proud of our achievement. Once stuffed, the driver would carefully close the passenger door, avoiding any shoulders and limbs, climb in, then drive. I have memories of ELO belting out of the tiny bass-blown speakers, someone's head wedged into my back, a bony elbow in my ribs, and my face pressed up against the window, eyes wide, absorbing the fast flicker of hedges in full colour as we hared along those narrow lanes to the beach.

Usually the packing was so dense that invariably none of those sitting in the front passenger seat was in a position to extricate a hand and open the door. So, on arrival, we'd all have to wait for the driver to get out and open it for us. Like some multi-legged insect, we'd pour out of the two doors, to unravel into separate entities sporting white limbs, spotty faces and dazzlingly inappropriately sloganned T-shirts.

A handful of us had wetsuits, but most didn't. None of us really registered the temperature of the water. Armed with boards, we'd stroll down to the foamy edge left by the last retreating wave, have a vague discussion about how long we'd aim to surf for, and then wade in. I think we all understood that any agreed time frame was always elastic… one hour could easily stretch to three or four, depending on how hungry, cold or churned up we became.

When I tell people that we used to bodyboard, many seem surprised, disappointed even, that I didn't learn to get up off my knees and surf 'properly'. And of course, now I do regret that I didn't, but it wasn't really on my radar then. That pick up and surge forward, body pressed close to wood as it carved great wings of water either side of my ears… peering around to see if any of the others might've caught the same wave. The loony grin shared when we'd spot each other. The sense of speed was intoxicating, and addictive. Even if our fingers and toes began to lock up and turn blue, we'd still be desperate to catch just one more wave.

As we became older, the group expanded with friends that became family too, and we tended to head towards another hangout, one that required a lengthy ramble along the coast to a faint, zigzag trail running down a vertical mass of loosely tethered grass and bramble. Almost invisible to the casual observer, the only way you could see it was to look for the pale purple haze of sea scabious that tended to grow along the edges of the path. We knew about it because our parents used to clamber down the same vertiginous track as teenagers themselves. And a little later, they'd climbed down with us as babies, strapped to their backs while free hands held on to picnic bags and clumps of grass. Access to this tiny stretch of sand is nigh on impossible unless you know of this path, have a boat or wait until the tide is fully out so you can walk around from a neighbouring cove. We rarely waited for the ebbing tide though, since at that point 'our' beach would be 'invaded' by others. This little bay is just a long wedge of fine sand and a vast, perfectly angled slab of smooth slate that on a clear day would absorb the sun as she worked her way around the coast. We'd stretch out on towels, open a paperback and roast. It horrifies me now to think of the measures we took to develop a tan. Sun cream had very little to do with it. Turning lobster-red was seen as a major achievement and one step closer to being caramel. If we became too hot, then a quick dip in the sea was the answer, rather than getting out of the sun's gaze! Some of these dips might be lengthened if a ball was involved, or if perhaps a swarm of silvered young herring was found. At which point all thoughts of tanning would be abandoned and we would pile in to see. Any sea life would interrupt tanning. But just sometimes we'd be visited by a Portuguese man-of-war, at which point a dip might have been shortened. Some of us, however, felt drawn to see these infamously dangerous, yet exotic, creatures for ourselves. I remember one year when their arrival even made the national news. Seemingly, fleets of these ocean-feeding, colonial organisms had drifted here on westerlies to enjoy the waters of the north coast of Cornwall. With metres of long, violet-indigo tentacles suspended from their transparent, rainbow-dipped, inflated air sacs, they reminded me of spent party poppers. One cousin decided that flippers and a snorkel were in order, and at the time I recall treading water alongside one of these 'floating terrors', marvelling at its myriad of colours, while he dived down to investigate its 'legs'. Thinking back, it was a daft thing to do. But he bobbed back up to smile, gulp air and then return. It's miraculous that he didn't get stung that day.

I look back and can see that, as teenagers, for all our efforts to engender the cool and beautiful stereotypes that we admired, it would have taken very little for the façade to crack and the child to emerge with an honest fascination and enthusiasm to explore and learn more of what this sea had to offer.

We're all older now, most of us with families of our own. And although we are spread wide over this island, and further afield, we've all tried to introduce 'our' beaches to our offspring and keep this feeling of connection to land, sea and family alive and thriving. It's not been easy, and at times when we couldn't travel back due to rubbish finances or work commitments, I've felt torn and disloyal to my roots, my family and to our children.

However, even though we couldn't return to those beaches and cousins as much as we would've liked, somehow enough loose threads have been unwittingly gathered up, and woven into a tie strong enough that our children felt an attachment. As they've become older, they choose to make the journey themselves… to those beaches, to the sea and to connect with their kin. But more than that, they're seeking out, fostering and nurturing their own relationships with new parts of our coastline, and with all the inherent wildness to be found there, and they're sharing it with their friends.

I'm driving home from a long and drifting swim in a little cove that hasn't yet been discovered by the masses that come to stay along this little stretch of coast. Inevitably, the pockets of my shorts are bulging with new treasures, among them striped stones (humbugs, as my grandfather used to call them) and limpets that dig into the crease of my hip. Things to study, to draw, to hold. The red valerian is nodding at me, I wave back. It would feel foolish not to.

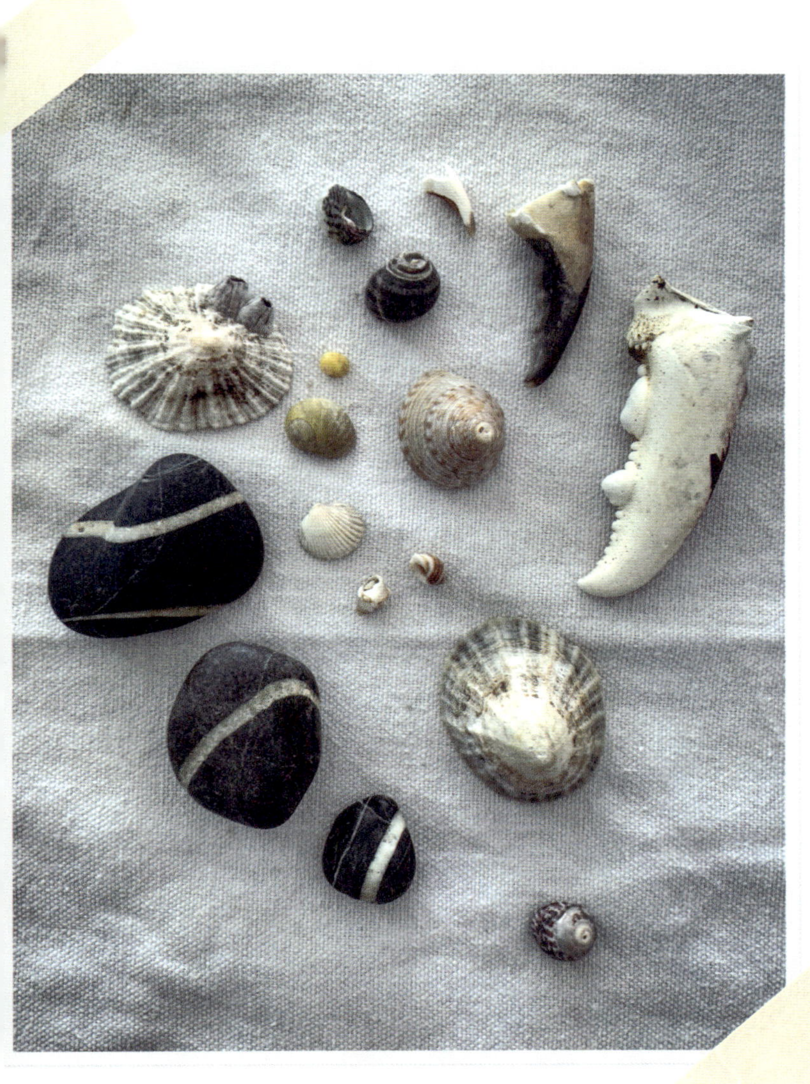

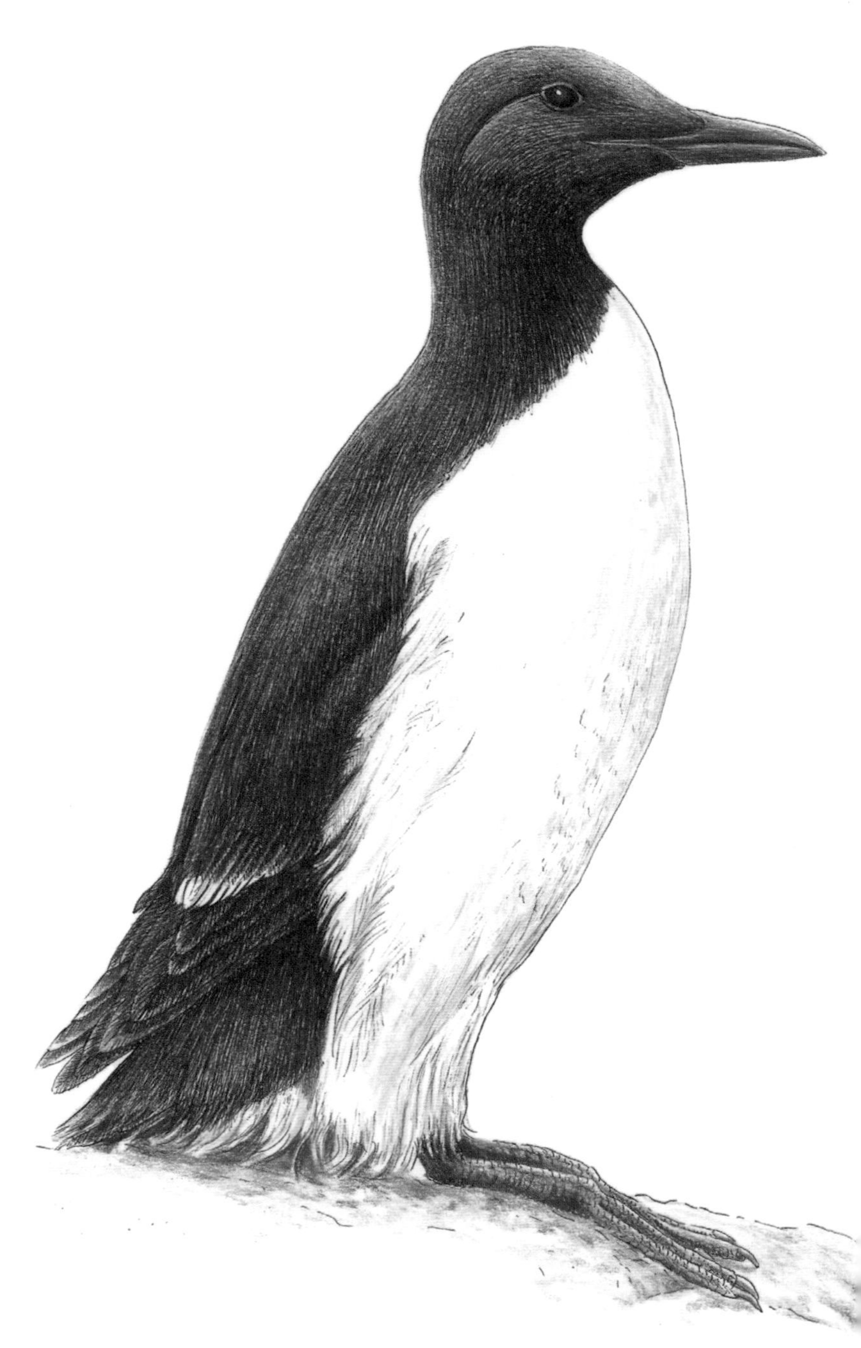

# Diving In

Using their strong, short wings alone to propel themselves, guillemots can dive and swim at high speed to depths of up to sixty metres in search of food. It was quite the surprise, though, when one was caught on camera at ninety metres in the North Sea by a submarine. Even employing superhuman efforts, and an oxygen tank, this would leave us about seventy metres short and scrabbling for the handle of the decompression chamber while the world turned sooty black. Scientists aren't entirely sure just how guillemots survive these deep dives. It's theorized that they absorb extra gases into their bones' vascular structure. Once cresting the water, the gases are then slowly released from their bones back into the body, avoiding the lung collapse and diving sickness that we humans would undoubtedly suffer.

We're fast approaching the end of June. It's early in the morning, and I am bobbing about among the lilt and drop of some gentle waves, keeping my feet busy, and shuddering a little at the thought of the few metres below me now, and what they might hold. I'll keep pedalling. If my feet keep moving, then surely, they're not an easy target. Wonderful logic there. Even after all the time I've spent in the sea over the last nine months, I have an annoyingly overactive imagination when it comes to monsters. But I'm here now and trying to imagine what it would be like to have a guillemot bob up beside me, fish in mouth, then taking off and flying to its nest, to feed its young chick. It's unlikely to happen, not only because I'm swimming in entirely the wrong place for them to dive (too shallow in this bay) but because all the guillemots along these steep cliffs will now be preparing to leave their crowded nesting sites, and take their weeks-old chicks out to sea. Usually, the male will stay with his offspring for around two months as they swim away from the coast and any predators. My thoughts are now drifting towards the most extraordinarily bizarre and dangerous aspect of their parenting, which equates to pushing a barely crawling baby on to a main road to see if it can make it safely to the other side. But more on that later.

Since we moved here, I've rarely left the house without binoculars, always with the hope of spotting new birdlife. Of all the seafaring birds that I've read about, and the few I've been lucky enough to see, the guillemot has to be my favourite, not only for its undeniably handsome attire but the sheer gutsiness with which it approaches its life and the job of childrearing.

I saw them for the first time on a chill day in early winter, while standing on the beach wrapped up against the cold and looking out to sea. Far beyond the break a string of perhaps half a dozen were flying, fast and straight, no more than a metre above the water. They kept their short dark bodies braced, their wings beating rapidly to maintain their pace over the waves. I hadn't really registered what they were; it was only later when I played it back that it became obvious.

Then, in early May, I caught a whisper that we have a very active colony of guillemots, just off the coast here. I began to do more reading about them and got in touch with a local guide to fix a date to visit the colony.

Dodging the grumbling rain clouds a couple of weeks later, I took a ferry ride to the other side of the Dart and drove out to the nature reserve beyond to meet up with my guide. We walked a short way while he explained a little about the site and its other feathered visitors. Hoping we might get lucky and see some of them, I made a mental note to write them all down.

We settled in the hide's lee, protected a little from the driving mizzle. My gaze was drawn across the water towards the sheer, craggy face opposite. Crammed like notes on a badly drawn stave, perhaps more than a thousand guillemots were packed in there, nesting wing to wing along the narrow shelves of the vast limestone cliffs.

Guillemots spend most of their lives far out at sea, only heading to land during the breeding season, from as early as March until the end of June. Apparently, this colony is slightly different. Due to the milder conditions here on the south coast (and perhaps a healthy supply of sprats and sand eels) they tend to hang around just outside the bay until October, when they make their way back to their ledges. Interspersed among this bustling colony there were clusters of razorbills, and a large gathering of kittiwakes.

I couldn't stop staring. Eyes back on the guillemots, I could make out singletons, interlopers, little spats occurring here and there, and couples. I locked on to one pair. They were going through a strangely familiar routine; child-rearing duties were being negotiated. Invisible to my eye from this distance, I presumed an egg was being shuffled from parent to parent, freeing up the other to head out, in search of food; a guillemot will rarely, if ever, leave its egg uncovered. Swinging the focus along a ledge and down a little we spotted a lone great black-backed gull sitting quietly among the colony. Guillemots have a few notable predators, herring gulls and carrion crows among them. This gull has apparently devised a rather cunning tactic to ensure his daily food quota is met. He's been spotted on more than one occasion shocking an adult guillemot into opening its wings in defence, whereupon he grabs one of the proffered wings and pulls the guillemot down into the water, to drown and eat it. A clever gull, no doubt about it. As I get older, however, I find it increasingly difficult to remain objective about this kind of underhand and calculating behaviour. For me it's on a par with the mob of magpies I used to watch when living in Sussex. They'd work as a team,

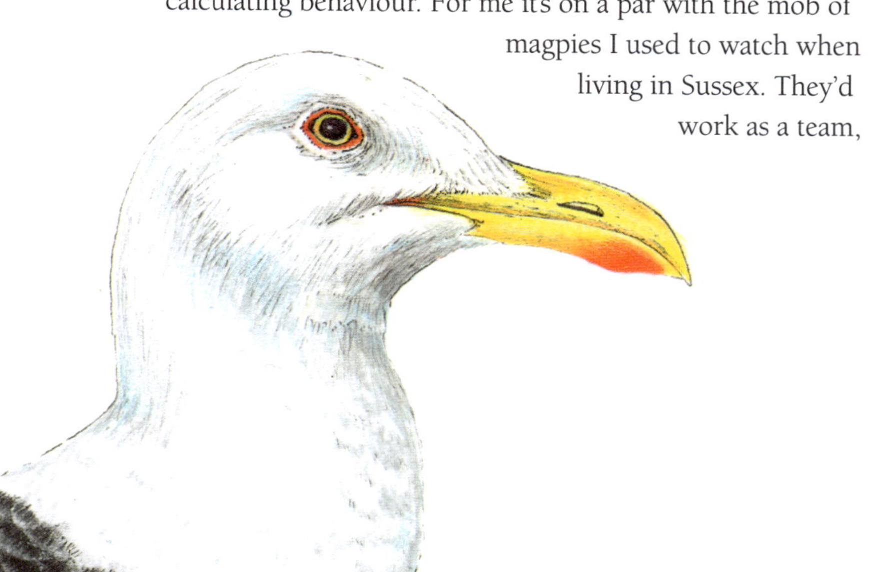

mock-threatening nesting birds to distract them, render them defensive, and ultimately draw them away from their brood, so another gang member could nip in and steal the vulnerable contents of the nest. Thankfully I wasn't the only one struggling. My guide's face was a mirror of mine.

Back from my swim. Skin salty and hair still dripping, I'm pawing over the pages of an egg identification book. With a variable base colour that can range from Mediterranean turquoise and dried-grass green to clotted cream and burnt sienna, guillemots' eggs are large and pyriform, and often highly decorated with wild and abstract markings that range from Miró to Pollock in their flair and design. Apparently, each egg's artwork is individual and unique to the mother bird, thus avoiding confusion among the crowds of tightly packed and presumably exhausted parents. Theories range on the reasoning behind the egg's pear shape, with most believing it evolved to avoid the tragedy of rolling off a precipitous nesting site. Another guess is that this shape makes it tough, so guarding it against possible breakage in the event of its being sat on or knocked about during a scuffle between neighbouring birds. Either way, just looking at the variety of patterns leaves me in awe. The only other eggs I know of that display such marked diversity are those of the skylark.

Since that visit to the colony in May, the piles of reference books I've been amassing seem to have taken up permanent residence like a Manhattan skyline closing in on my computer; an ever-growing bristle of lurid green sticky notes have been slapped on the wall, and the countless open tabs on my computer offer up a fairly accurate representation of how my mind feels right now.

However, all I can think about is that final leap of faith made by the chick at barely a month old. Rather sweetly, these young adventurers are called 'jumplings', which somehow belies the very real danger this rite of passage entails.

You may have seen this moment captured in a video and posted online. It will leave your stomach in a clench of knots. The parent birds, having fed and nurtured their one and only chick for that year, decide that the time has come for their stubby-winged darling to leave the ledge, the only home this little mite has known, and drop to the sea. Unfortunately, if you've seen the footage as I have, their fall is sometimes broken by the occasional upsurge of rock. This can spell disaster with the young guillemot not recovering. But, quite astonishingly, this is not always the case. If it has enough feathers for cushioning, this flightless morsel can bounce, gather itself together and then continue its obstacle course to the water, where its father will be waiting to welcome it to its new 'home'.

It's early evening now. The day has been overcast, but warm, the sea calm (so I'm told, since, having returned from the swim, I've been bent over my desk, illustrating eggs). I grab my binoculars and a sweater, and drive to take the ferry back to the guillemot cliff. I've been given the nod that this evening might be 'jumpling time'. As I arrive, there's already a small gathering of other enthusiasts. One colony watching another. We're all politely distanced along the edge, just a small strip of red valerian and viper's bugloss between us and the possibility of a few human jumplings if we're not careful.

I'm pulling on my sweater, and someone shouts that they've seen a chick falling. Head out, forcing arms through sleeves, and I'm grabbing at my binoculars and swinging them towards the sea. There's not a breath of wind and I can hear the wild chatter and call from the flotilla of parent birds floating on the water, as they urge their chicks to jump. And oh my god, they're jumping. They're really jumping.

It's a sight that I don't think I will ever forget. The blind faith of the chick… to push off and plummet some sixty metres or more. The absolute certainty of the parent, that this is the right way, the only way. It's seems desperate. The willingness to gamble absolutely everything in that one flightless moment is shocking. And yet, astonishingly, each one that I manage to track hits the water and immediately starts pedalling its feet towards an adult bird, 'Hey dad! I did it!' One even immediately decides to have a go at diving, even though it's never done so before, bobbing up to then paddle furiously across to the waiting parent.

Blind faith. It's something I'd always railed against as a young adult… Always demanding proof that something is so rather than just hoping it might be. But with three young adults of our own now, I have come to understand that much of what we have achieved together as parents has often been built upon hope and a belief that things will work out. We risk everything every day as parents. We're not pushing our children off a cliff – that would be daft – but everything we do is geared towards them leaving the safety of the nest. The first time they sleep on their own; the first time they go swimming without armbands; the crossing of a road without you; the first time they go to stay with friends; when they pass their test and drive away in a car.

And all we can do is trust that they will survive and thrive to go forth and do the very same that we've done. Insane?

Perhaps not, just hopeful.

The light is beginning to fade a little now and the cliff is becoming quieter. The show is over for today. Time to head home.

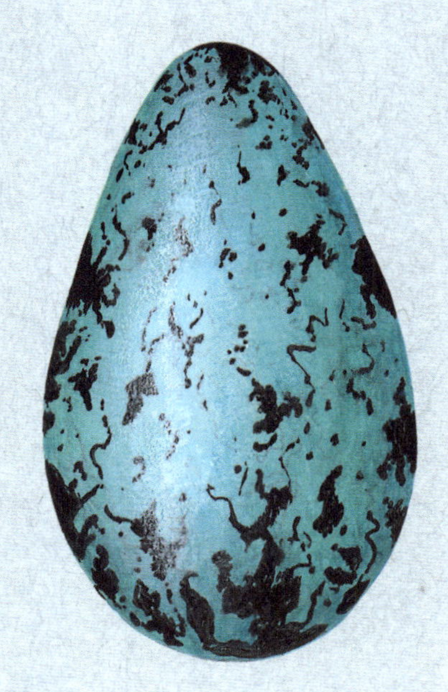

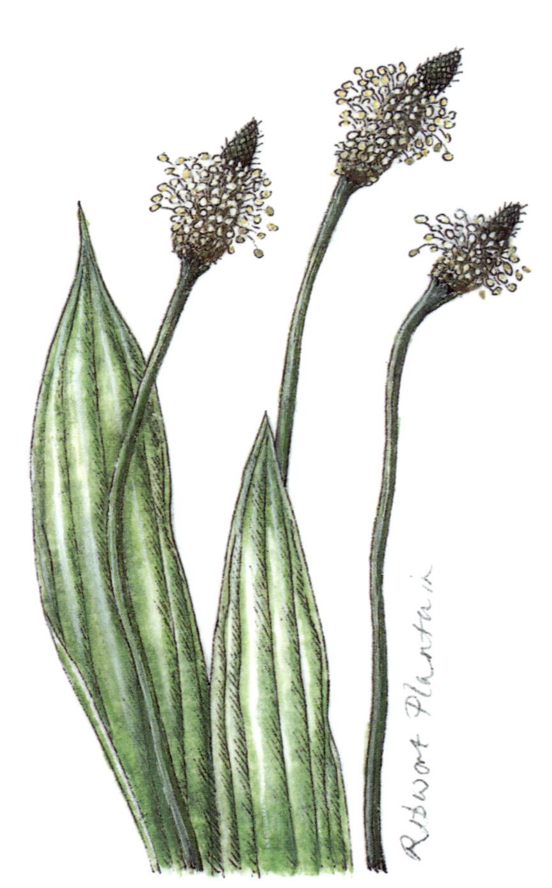
Ribwort Plantain

*10 July*

*Warm today! Climbed the field at the back*
*of our house.*

*Views out and across a valley of folded fields.*
*The song of skylarks above is caught and brought*
*closer by the wind.*

*More than one hundred years of sheep footfall*
*have terraced this steep rise; settled in the*
*ridges there's wild carrot, a scatter of mallow*
*and so many ribwort plantain, like ballerinas*
*in their tutus.*

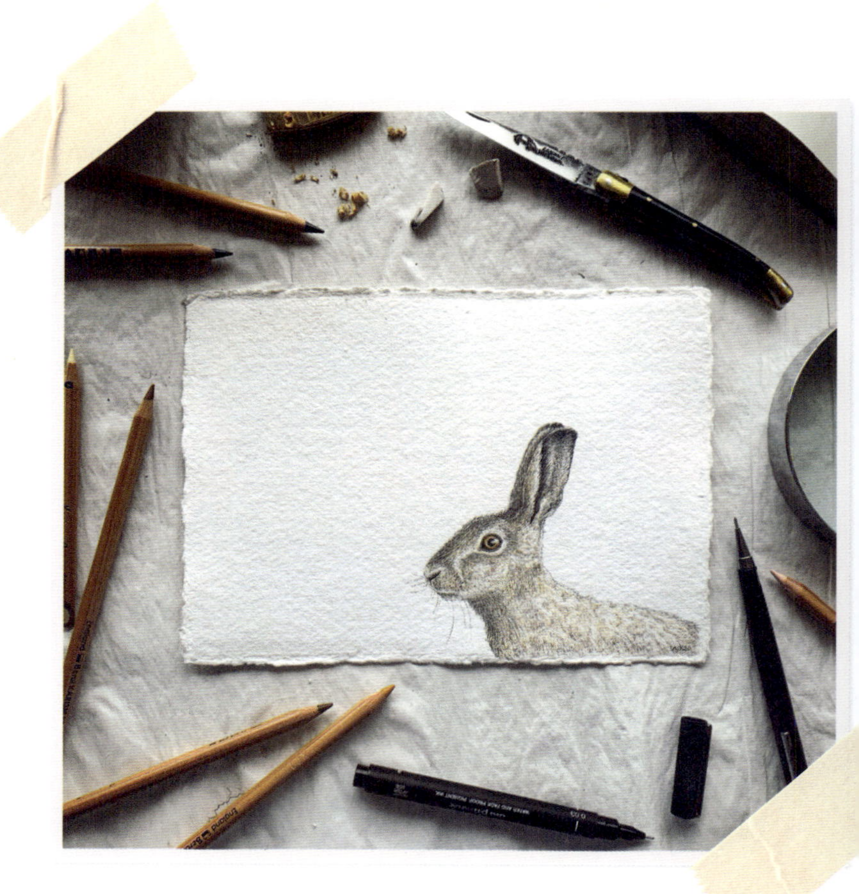

# Hare Here

An early start because the song thrush insisted. His gentle but unrelenting voice has managed to unearth me from a deep sleep, in which I dreamt of a brown hare, of all things. I would have very much liked to stay there and draw out the story into a conversation between us, the hare and me.

It's late July, and although our bedroom window faces north and the curtains are pressed tight, the morning sun still manages to slip in around the edges to illuminate the rumpled duvet and my half-hidden face. She's clearly conspiring with the song thrush. But I'm up now. And instead of reaching for a dressing gown, I've shrugged on some clothes and crept downstairs. I've been doing this a lot recently. Sometimes I'll grab a towel and a bikini, a flask of tea if I'm organized. Then I'll jump in the car and head to the coast. Swapping dressing gown and wellies for bikini and towel: I think this is slowly becoming my new thing.

This is also a bit of a tactical move, since we're now climbing towards peak holiday season and the skinny roads are beginning to choke up a little as the morning progresses. As are the beaches.

I've long had a thing about hares. I think they first made a meaningful appearance when I was about twelve and we were given a picture book, *Masquerade*, by the artist Kit Williams. Bursting out from among the pages were the most enchanting illustrations. These were paired with deceptively familiar prose, taken from well-known nursery rhymes and repurposed to become riddles for a beguiling treasure hunt. The clues were woven among the intricate paintings, and the prize was an 18-carat golden hare. I'd never considered the hare and I didn't really feel compelled to try to unravel the many riddles. I just couldn't draw my eyes away from the illustrations. They were richly portrayed artworks of nature, sometimes with humans included, but always with nature in charge. I found them unearthly, untethered and a little dark… just enough to ignite an unease within me. And it was the hare that teased and unsettled most. That innocent gaze with a hint of humour in his eye. I loved the book, and I was a little scared of the hare.

I don't recall many encounters with hares – there were plenty of rabbits, both bounding across fields and contained within one of my mother's 'chicken stews' – except for one. This was when we looked after a tiny foundling that my father had spotted on the side of a road. Weak and malnourished, this little mite didn't really stand a chance. We named him Harvey, cared for him and eventually put him in our chicken run. He settled in happily and would roost and eat with the girls and sit among them while they preened and fluffed in their dust-baths. He was let out and free to roam, but would always return, come the evening. He was so certain that he was one of their kind, that the introduction of a cockerel resulted in a major boxing match every time the poor bird endeavoured to approach one of the hens to fulfil his role.

Then one day Harvey went roaming and didn't return. Perhaps he found a mate, and with a new perspective (and new wife) re-evaluated his role in the coop. We never saw him again.

My grandfather firmly believed Harvey to be a brown hare. My parents were less certain. But if indeed he was, then to be domesticated, and ultimately reduced from a wild thing to a shadow of its species would've been wholly unjust. That he was still hard-wired to roam, and ultimately leave, is heartening.

Over the years, I've read books and articles about hares; it's been a very gradual education. Due to their purported weakness for tender plants, they're viewed as a pest by many farmers, although in truth they prefer wild grasses and herbs. But with the continued degradation and often complete removal of hedges and their margins, this leaves them little choice but to turn to crops for sustenance. I've learnt that they're the fastest land mammal in Britain, hurtling at speeds of more than 70 kilometres per hour. The only thing that can outrun them is a greyhound, or a bullet. I've also learnt that their numbers are dwindling, with the brown hare declining faster than any other native wild mammal, apart from, perhaps, the water vole. And then there's vast quantities of literature on the more mystical, magical nature of hares; their rise through history to prominence as portents of good luck, health, speed and vitality has been amply documented.

Until a few years ago, I'd never seen a hare in the wild. Then one summer, while driving at day's end along a dusty road in rural Tuscany, we passed by a stretch of fallow land, and there they were. Three of them, paused and alert, their ears backlit and made transparent by the retreating sun. Two more pairs of sooty-tipped spears appeared, followed by their owners. Windows wound down, we slowed to a stop to watch. These five continued to stare back, heads turned slightly to allow their widely spaced eyes to focus. They viewed us with an air of faint interest; mouths chewing and an occasional ear flickering as if in mild irritation… a fly perhaps. Then suddenly a silent cue was shared, and they bolted, zigzagging towards the sprawling, shadows of the edgelands. I later found out that this tight and precise manoeuvre is called telemarking. With longer hind legs than a rabbit, and more powerful forelegs, hares can suddenly change direction, a useful tactic when being pursued by a predator. Within seconds the field was still once more, the sun's last gaze settling on the bleached heads of forgotten wheat, and baskets of wild carrot.

I've read that some have made their homes here, along the edges of the farmland near the coast, but that they're quite elusive, tending to be more visible at first and last light. So these early morning drives have begun to take on a new purpose: I'm hoping beyond all hope that I might see a brown hare.

In the car, driving. The sun is high enough to cast her eye along the lane ahead, and the shadows that owned this

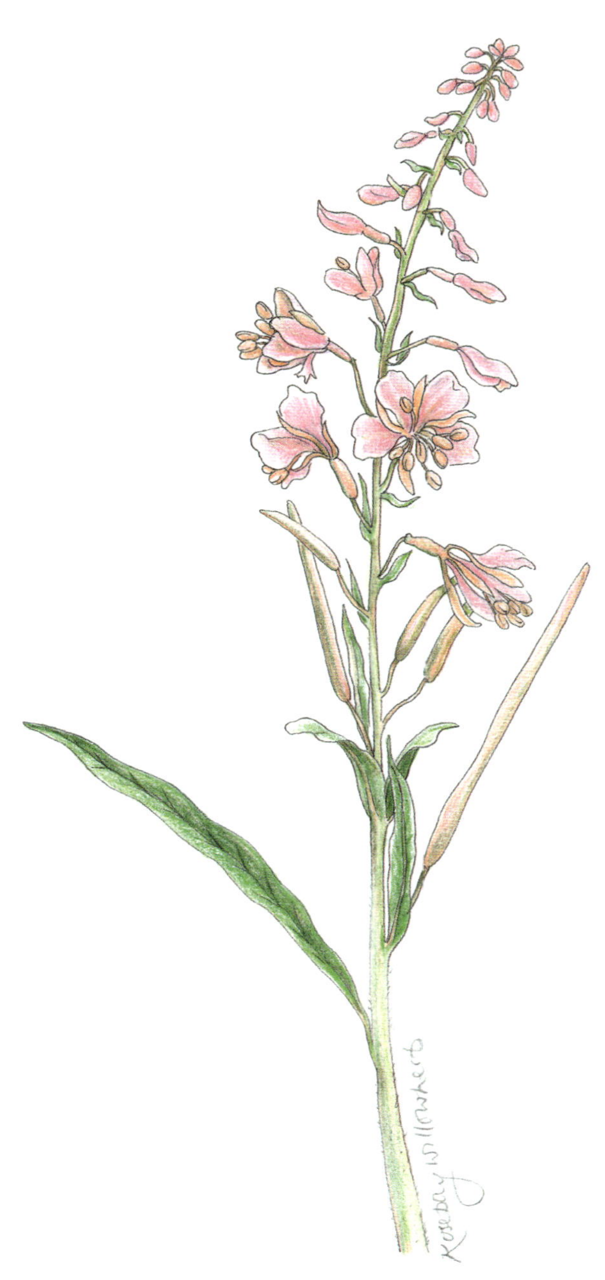
Rosebay willowherb

land just hours before are now reluctantly hugging tight to hedgerows and buildings, lest they be found and banished. The last of the cow parsley is beginning to topple as their leggy stems stagger beneath the abundance of umbrella-ed froth above. Like a crowd-shy actor pushed on to the stage, a hare stumbles out from a curtain of rosebay willowherb, pauses as if gathering his thoughts, then slowly trots along the road a few metres or so ahead of me. He seems not the least bit worried that I'm following him. In fact, his pace is almost leisurely. We continue this way for a few yards, passing the time. Then he steps up his pace into an easy lollop, slows once more, and finally steers off the road towards a small gap in the hedge and into the field beyond. I draw level with the leafy portal. Not a sign of him… just the burnished glow of the sun-ripened crop beyond. The temptation to get out of the car and squeeze through is making my heart drum.

I am whooping and cheering, the only member of a star-struck audience. There's no one with me to witness the moment, no one to turn to and say, 'did you see him?' I'm just over a kilometre from the coast now, but any thoughts of an early morning swim have been trumped as I pull into a lay-by, turn the car around, and head home to draw.

*1 August*

*Morning swim at Sugary Cove.*

*Sat and chattered awhile. Sand martins* (Riparia riparia) *came to cruise the high tide for breakfast!! A joy to have them back here.*

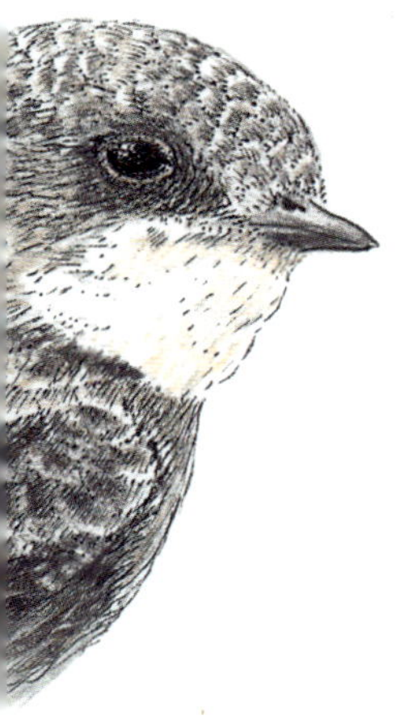

# Sea Calendar

Welcome August... a month when summer might consider that her work is done.

Having festooned hedges, headlands and field margins with an abundance of flora, and in so doing provided for the wildlife and all its endeavours to procreate, now would seem to be the time for her just to sit back and shine a while longer, ensuring that the rich and lusty ripening of all that was born can continue to its full and glorious potential.

However, we all know that quite often summer tends to clock off rather early, usually around the time when families have carefully crammed their kids, suitcases and perhaps a dog or two into their car to embark on their annual holiday.

Still, who am I to judge? It's early days yet and, so far, everything seems to be going to plan.

I'm walking along the cliff path towards one of my favourite coves. The last blooms of gorse have all but gone, and the pale clouds of bramble flower have been replaced by shiny clots of blackberry, so heavy that thorny arms now hang low… a little weary, but content to offer up their jewelled bounty. No doubt this will please the much-expanded family of stonechats that are still very much present along this headland. This morning, the sun has pushed her way through a tightly packed mob of thunder clouds. I'm not sure how she managed it, given that they were more than a little resistant to share the sky with her. However, she's now making the dry and whispering grasses gleam, and the sea beyond the stepped slate cliff a radiant flock of jostling, silvered slithers.

These months have sped by at such a rate that I've frequently found myself having to double check the date.

If I didn't have a calendar to glance at, then I'm almost certain that I could tell you the time of year by the colour of the sea. Often she seems to straddle two seasons in one stretch as if uncertain of her artistic direction, and so the palette will flex with such a myriad of hues that I am left struggling to catalogue it. But it's a happy struggle. Where once I would watch the hedgerows, fields and forests to understand and connect with each month, I instead note the colours of the water, mentally sifting through my collection of pigments in order to try to identify them, make them relatable and hopefully reproducible in paint. And, when that fails, I resort to making new names when there are none to be found.

This is not a new approach. In the early nineteenth century, the mineralogist Abraham Gottlob Werner set out a system for recognizing minerals by their key characteristics, including their colour. He would attribute to each mineral a colour and make a note of its base tint, then chart the different branches of that hue, being acutely aware that his descriptions could be entirely subjective. So, as well as the mineral (in this example, Beryl), he referenced animals ('egg of thrush') and – if applicable

– a corresponding vegetable ('under disk of wild rose leaves') to properly describe the essence of this colour in the natural world. A little later his scheme was adapted to include a corresponding chart, painted by Scottish flower artist and teacher, Patrick Syme. Only one hundred copies were published, with each containing meticulously hand-coloured squares to perfectly reproduce Syme's interpretation of Werner's original work. It was seen as a great success, and used widely by artists and naturalists alike, the most notable among them being Charles Darwin, who took a copy with him on his voyage aboard the HMS *Beagle*.

To an artist this book is more than just a charming artefact. The only other colour system of which I'm aware is the Pantone Matching System, devised in the 1950s by chemistry graduate, Lawrence Herbert. Although it offers a workable and standardized colour scheme that's recognized the world over, it is wholly lacking in that visceral connection that Werner created. His keenly observed and noted tints flow like lines of verse.

Each day that I have visited the sea, to stand eye to eye with her, she has shown herself to be a conjuror of colours. Irresolute and noncommittal in her choices, she seems to thrive on exploring every permutation to create new and previously unimagined shades. And yet, each season, she seems to have a recurring theme that reflects not only the temperature and the weather of that quarter but also its mood.

Looking down now towards the water, the sun's reign almost over as that troop of grey prepares to snuff her out, there's a hint of ultramarine with flecks of muted cadmium orange. However, the deep core of her swaying body is wearing the darkest of Prussian blues, stirred through with fleeting curls of inky purple. If Werner was standing beside me, he would have described this colour as that of the egg of the man-of-war; the purple auricula would be its likeness in the plant world and fluorspar (or fluorite) would be its mineral equivalent.

Dipping down to find the familiar gap along the bluff of blackthorn, I make my way through and descend to the little cove to swim in its shifting palette.

If I were to try to guess the mood reflected in the sea today, I would tell you that it is one of strength and calm. There's hope there too, among those specks of cadmium orange. Werner called this colour buff orange and likened it to the 'streak from the eye of the kingfisher'. As I swim among the folds of this summer-warmed sea, I'm recalling the kayak trip I made last winter, when I saw the azure-emerald flash of my first kingfisher as it flew fast and low across the dark cold river. It feels like a long time ago.

We've been living here, in this coombe near the sea, for almost a year now, and as friendly and welcoming as everyone has been, I'm acutely aware that we will probably always be seen as the newcomers, regardless of however many more years we live here. Yet every cove along the sea-sculpted edges of this coast has felt familiar and kindly.

And the ever-changing cloak of flora and fauna that flows from moor to sea has pulled me in to reveal new treasures to study, draw and love.

Just one year.

But I belong.

The sun has found another chink in the armour of this leaden bank of unspent rain. It's just enough to warm my back as I hop from one foot to another, endeavouring to balance and pull on jeans. Sitting down in the coarse sand would be easier, but I've learnt by now that I'd be taking back more beach than I'd planned on!

One last look, and she's changed again... swirling cobalt with a smudge of mallard's wing.

Time to head home and paint that last square of sea.

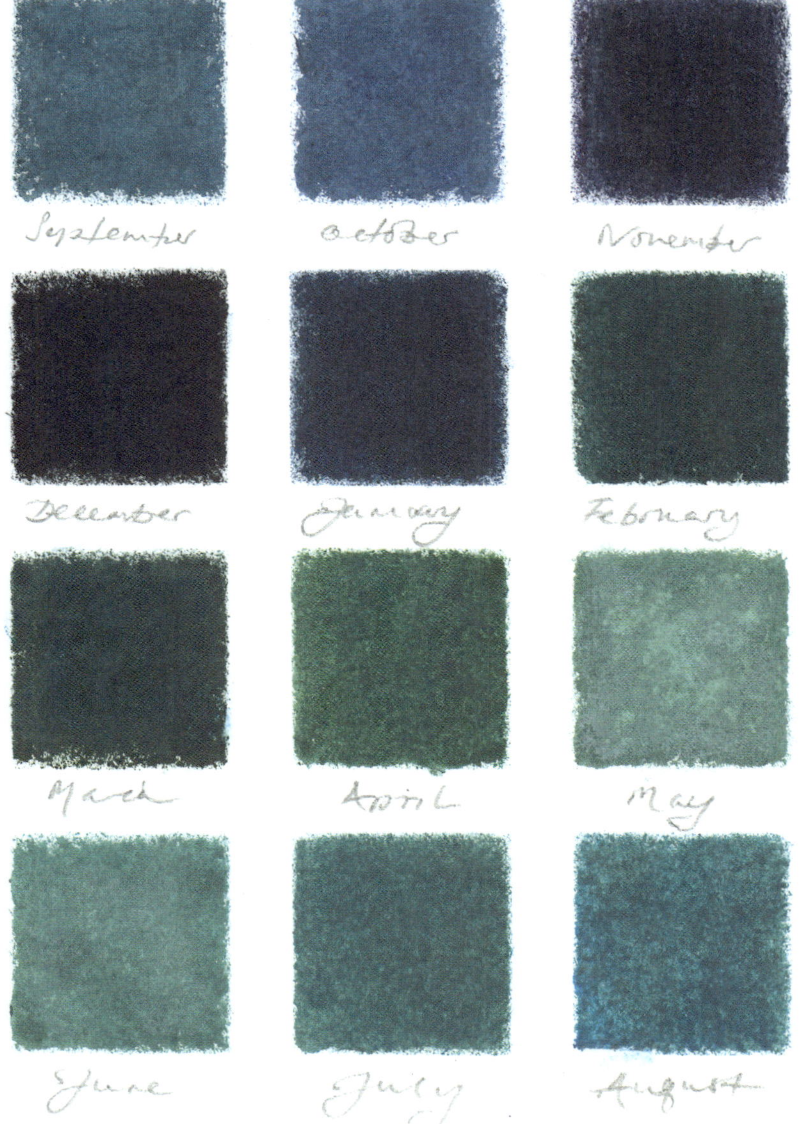
September
October
November
December
January
February
March
April
May
June
July
August

# Thank You

Dearest Koska, well I picked a fine time to write another book, didn't I? Just as we were unpacking to begin a new chapter, here by the coast.

Thank you for giving me the space and the time to explore, research, write and paint.

Flora, Ellie and Jolly, as ever you've been so supportive and enthusiastic. Thank you for letting me babble on, often unrelentingly, about the birds, the seaweed and the seals.

Dear Emily, your guiding hand throughout has been invaluable. I am forever grateful.

To all at Pavilion. Not being able to sit around a table together and plot and scheme this book into its final form has been a challenge at times, during this strange year. But you made it so much easier than it might've been. Thank you for your patience and commitment to making this book something to be proud of.

Mum and Dad, my constant source of inspiration. My love and thanks to you both.

Finally, I am beyond indebted to all of those who I've been lucky enough to meet and become friends with over the last year here in South Devon. You've been so welcoming and your kindness towards me and my family has made us feel settled, and very much at home.

PS To Billie, my constant companion: Shall we go for that walk now?

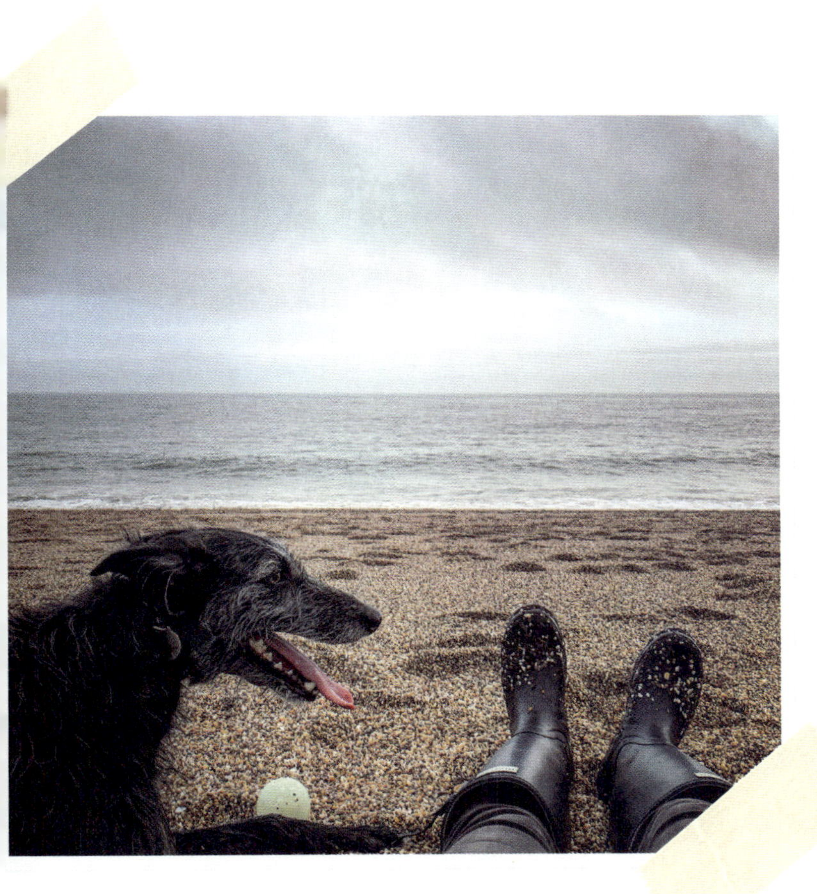

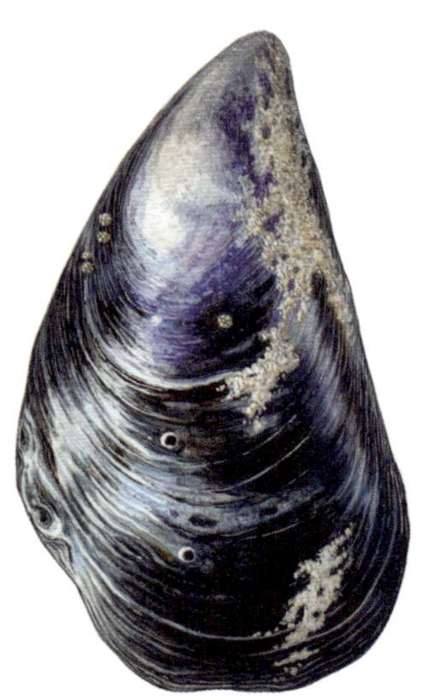